THE AMATEUR ASTRONOMER

SCIENTIFIC AMERICAN

THE AMATEUR ASTRONOMER

Edited by Shawn Carlson

JOHN WILEY & SONS, INC.

New York · Chichester · Weinheim · Brisbane · Singapore · Toronto

Published by John Wiley & Sons, Inc.
Published simultaneously in Canada

Illustrations 11-1 and 11-2 copyright © 1990 and 12-1, 25-1, and 26-1 copyright © 2001 by Michael Goodman

The publisher and the author have made every reasonable effort to ensure that the experiments and activities in this book are safe when conducted as instructed but assume no responsibility for any damage caused or sustained while performing the experiments or activities in the book. Parents, guardians, and/or teachers should supervise young readers who undertake the experiments and activities in this book.

Library of Congress Cataloging-in-Publication Data

The amateur astronomer / edited by Shawn Carlson.
    p. cm.
   At head of title: Scientific american.
   Includes bibliographical references and index.
   ISBN: 0-471-38282-5 (pbk. : acid-free paper)
    1. Astronomy—Amateurs' manuals. I. Title: Amateur astronomer. II. Carlson,
Shawn. III. Scientific american.

QB63 .A23 2001
520—dc21
                                                                 00-047773

10 9 8 7 6 5 4 3 2 1

CONTENTS

FOREWORD

The Amateur Scientist

The present volume, titled *The Amateur Astronomer*, is intended to be the first of a series of books that will derive from the longest running column in *Scientific American*, now and for the past 50 or so years known as "The Amateur Scientist." The origin of "The Amateur Scientist" goes back to a column first published in May 1928 and written by Albert G. Ingalls. It was called then "The Back Yard Astronomer." His first sentence in that new column declared: "Here we amateur telescope makers are, more than 3000 of us, gathered together in our own back yard at last." At the top of the page is an illustration of the "Back Yard," with an amateur astronomer at work, apparently drawn by one Russell W. Porter, then considered a mentor of telescope makers throughout the land. Porter declared that the name of the new column, contracted to "Backyard Astronomer," conferred an honorary B.A. degree on all its readers.

In April 1952 the column extended its coverage to include the work of amateurs in all branches of science, in addition to telescope making and astronomy. As a fisherman, one of my favorite pieces in July of that year is on the May fly, to which I have referred a number of frustrated participants in the sport. Although in 1952 it was virtually impossible to determine how many people spent leisure time in the pursuit of scientific study and experiment, one survey in Philadelphia determined that 8000 laymen in that city, known for some of Benjamin Franklin's most significant scientific discoveries, were actively engaged in amateur science, and that 700 of the group had made contributions important enough to warrant attention from various scientific professionals.

The column continued under Mr. Ingalls' leadership (and under several titles mostly related to astronomy) for 27 years, until his retirement in May 1955, at which time responsibility for the column was turned over to

C. L. Stong who continued to widen the breadth and depth of coverage. Stong died in 1976, and a new columnist, Jearl Walker, was appointed. As discussed in the following Introduction by Shawn Carlson, the character of the column changed under the editorship of Walker to focus more on understanding the principles of physics, his own field, as demonstrated in everyday life and less on interesting scientific devices within the grasp of the amateur scientist. Walker retired in 1990, and after a period of discontinuous publication, "The Amateur Scientist" was turned over to Shawn Carlson in 1995.

Happily to all of us at the magazine it is once again attracting an increasing readership of informed and interested "amateurs," who can now supplement their reading of the column and activities stemming therefrom with an increasing number of Web sites, including our own (*www.sciam.com*). Dr. Carlson has our heartfelt gratitude for putting "The Amateur Scientist" back on course.

I am often asked about the readership of *Scientific American*—who these readers are and where their interests lie. My strong suspicion is that most of these inquiries come from potential readers who for one reason or another have not yet made the leap. Summarizing our readership in a brief reply is very difficult, because the population is very large and very diverse and includes secret clusters of people who would not normally be identified as amateur scientists. If pressed for a brief reply, however, I often mention that my image of a reader of *Scientific American* is someone whose favorite fantasy on a rainy Saturday afternoon is to read "The Amateur Scientist" and to plan a future experiment.

John J. Hanley
Chairman Emeritus
Scientific American, Inc.

INTRODUCTION

Scientific American magazine first came off the presses back in 1845. That makes it the oldest continuously published magazine in the United States. And "The Amateur Scientist" holds the distinction of being the magazine's longest running column; it traces its pedigree back over 70 years. That surprises most folks. The longest running column in the oldest magazine in the United States isn't dedicated to concerns like sex, or scandal, or style. It's about good ol' Yankee ingenuity. It's devoted to helping everyday people explore their world and giving them a fighting chance to make original discoveries. That fact never ceases to give me hope for our troubled times. And it is my great honor to write "The Amateur Scientist" today.

Over the years, "The Amateur Scientist" has made a real contribution to our world. Many working experimentalists first kindled their interest in science while carrying out a research project described there. And generations of science teachers have relied on the column for exciting projects to challenge even their most gifted students. For all their discoveries, and for helping them spread a passion for science amongst young people, "The Amateur Scientist" deserves at least a little credit.

You might be surprised to learn that despite the column's influence and popularity for more than seven decades, only one anthology of its projects has ever been published. That book, *Science Projects for the Amateur Scientist*, which was edited by C. L. Stong, went out of print in 1972. Since then, that title has become quite sought-after. A copy in good condition now sells for well over $100, if you can find one at all. Clearly, it's high time some of these classic articles became more accessible to the amateur community.

It's fitting that the first new collection of these articles in book form be devoted to astronomy. Indeed, astronomy owes a great debt to its amateur explorers. Today thousands of dedicated avocational scientists scrutinize the night sky with extremely sophisticated instruments. Every year this largely disorganized rabble makes important discoveries that advance our understanding of the solar system, our galaxy, and even the cosmos. In fact, the amateur community boasts such outstanding scientific talent that some professional astronomers now partner with amateur observers, and the number of these collaborations is growing every year.

I don't believe that any of this would have happened without *Scientific American*. In 1928, a young astronomer named Albert G. Ingalls launched a new feature in that magazine titled "The Back Yard Astronomer." In the nearly three decades that followed, Ingalls' column changed names several times, ultimately becoming "The Amateur Scientist." During Ingall's tenure his writings (including his landmark *Amateur Telescope Making* book trilogy) as well as his tireless community-based efforts ignited the public's passion for hands-on astronomy. Today's community of amateur astronomers grew out of his efforts. These early works may fairly be said to be the founding documents of amateur astronomy.

Ingalls retired in 1955 and turned "The Amateur Scientist" over to C. L. Stong, a brilliant electrical engineer whose passion for science ran the gamut from astronomy to zoology. Under Stong's tenure the column expanded into many other fields, but it still kept in touch with its roots in astronomy. Stong published some designs for extremely sophisticated astronomical instruments, including devices to measure the chemical composition of stars, to view solar prominences, and to enhance the resolution of deep sky photography. The professional versions of these instruments are so advanced that today's amateur still cannot afford them. But Stong's expositions made it possible for amateurs to build these instruments inexpensively in home workshops, using ordinary shop tools.

After Stong died in 1976 from lung cancer, *Scientific American* hired a new columnist named Jearl Walker, a physicist and well-known science writer. Walker wrote many excellent articles, but under his stewardship the column was devoted more to interesting physics of the everyday world than to instrument building. As a result, "The Amateur Scientist" stopped featuring cutting-edge devices for which it had become so well known. This disappointed amateur astronomers in particular, who for two generations had come to rely on *Scientific American* as a great source of inspiration for their own research. After Walker retired in 1990, the feature drifted without a regular columnist, with articles appearing about every other month. Finally, in 1995, *Scientific American* gave me the nod to take over the feature, and I have delighted in returning the column's focus to hands-on science. That's why this compendium sports just one article from between 1977 and 1995, but several articles from the last few years.

A lot of the articles you'll find here are therefore old: some first appeared in the 1950's. So selecting and editing this collection turned out to be quite a challenge. You'll find every type of astronomical instrument that has ever appeared in "The Amateur Scientist" described here. Many of the earlier articles have great historical interest for the enlightened technophile. I often chose accessibility over history, and simply updated the text directly. But some articles gave such insights into how our predecessors approached important research problems that I decided to add sidebars. Occasionally I even threw in a brief introduction to set things in perspective. By taking this middle-of-the-road approach, I am quite certain that I have fully satisfied no one. Still, I hope the long-time fans of "The Amateur Scientist" will find this work to be a reasonable compromise between utility and history.

The amateur community owes a debt of gratitude to several people for making this present compendium possible. First, I must thank Diane McGarvey of *Scientific American* magazine, for being the irresistible force behind this project. Without her tireless efforts, this volume would not have been possible. Also, my darling wife Michelle Tetreault deserves much credit both for assembling the glossary and for the saint-like patience she has shown me while this book was being cobbled together. And lastly to my children Katherine (age two) and Erik (age four months) for teaching me about what really matters in this world and for making me so grateful each day for being a part of it.

THE AMATEUR ASTRONOMER

PART 1

TELESCOPE MAKING

1 A SIMPLE TELESCOPE FOR BEGINNERS

Adapted from C. L. Stong columns, November 1959
October 1969
December 1955

Back in 1925, an article in *Scientific American* described how a group of amateurs in Springfield, Vt., made a reflecting telescope powerful enough to show the mountains of the moon, the rings of Saturn, the Great Nebula in Andromeda and comparable astronomical objects. According to the article the instrument could be inexpensively duplicated by anyone willing to invest a few hours of labor. The details of construction had been worked out by Russell W. Porter, engineer and explorer, and were described in collaboration with the late Albert G. Ingalls, an editor of *Scientific American*. Within a year some 500 laymen had completed similar telescopes and were well on their way to becoming amateur astronomers.

I was one of them. Like many laymen I had wanted to see astronomical objects close-up, but could not afford a ready-made telescope of adequate power. Nor was I acquainted with the owner of one. The description of the Springfield telescope solved the problem. I immediately set out to make a six-inch instrument, and I had scarcely begun to use it when half a dozen of my neighbors started telescopes of their own.

It was not a very good instrument by the standards of present-day amateurs, but it showed the markings of Jupiter and the polar caps of Mars. The fact that scattered light gave the field of view a bluish cast which tended to wash out the contrast, and that the stars wore curious little tails, detracted not a bit from the satisfaction of observing. So far as I knew this was the normal appearance of the sky when it is viewed through a telescope! Over the years I made and used better instruments, and on one occasion I even enjoyed a turn at the eyepiece of the 60-inch reflector on Mount Wilson. By then, however, I had found observing almost routine. Even the Mount Wilson experience did not give me the same thrill as that first squint through my crude six-incher.

In my opinion the beginner should not attempt to make a really good telescope on the first try. Too many who do grow discouraged and abandon the project in midstream. The application of the tests and figuring techniques through which the surface of the principal mirror is brought to optical perfection is a fine art that is mastered by few. I have made more than 50 mirrors and have yet to polish a glass with a perfect figure to the very edge. For all but the most talented opticians neither the tests nor the techniques are exact. After misinterpreting test patterns and misapplying figuring techniques for some months the beginner is tempted to give up the project as impossible and discard a mirror that would operate beautifully if used. Conversely, spurious test-effects have been known to trick veteran amateurs into turning out crude mirrors by the score under the prideful illusion that each was perfect. That such mirrors work satisfactorily is a tribute to the marvelous accommodation of the eye and to lack of discrimination on the part of the observer.

Beginners may nonetheless undertake the construction of a reflecting telescope with every expectation of success. Amateurs with enough strength and mechanical ability to grind two blocks of glass together will be rewarded by an instrument far superior to that used by Galileo. They need not concern themselves either with tests or elusive figuring techniques.

The simplest reflecting telescope consists of four major subassemblies: an objective mirror which collects light and reflects it to a focus, a flat diagonal mirror which bends the focused rays at a right angle so that the image can be observed without obstructing the incoming light, a magnifying lens or eyepiece through which the image is examined, and a movable framework or mounting which supports the optical elements in alignment and trains them on the sky. About half the cost of the finished telescope, both in money and in labor, is represented by the objective mirror.

The mounting can be made by almost any combination of materials that chances to be handy: wood, pipe, sheet metal, discarded machine parts and so on, depending upon the resourcefulness and fancy of the builder. The mounting designed by Roger Hayward, illustrated on the next page, is representative. The dimensions may be varied according to the requirements of construction.

Materials for the objective and diagonal mirrors are available in kit form from dealers in optical supplies (see supplier list on page 255). Ama-

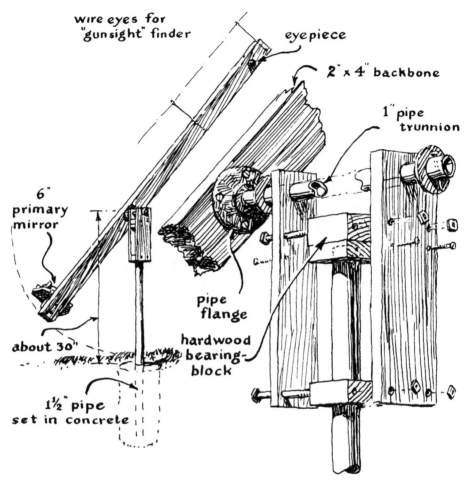

wire eyes for
"gunsight" finder

eyepiece

2" x 4" backbone

1" pipe
trunnion

6"
primary
mirror

pipe
flange

hardwood
bearing-
block

about 30"

1½" pipe
set in concrete

Figure 1.1 The mounting of a simple reflecting telescope

teurs with access to machine tools can also make the required eyepieces. The construction is rather tedious, however, and ready-made eyepieces are so inexpensive that few amateurs bother to make their own.

The beginner is urged to start with a six-inch mirror. Those of smaller size do not perform well unless they are skillfully made, and the difficulty of handling larger ones increases disproportionately. Kits for six-inch mirrors include two thick glass "blanks," one for the objective mirror and one (called the tool) on which the mirror is ground. The kits also supply a small rectangle of flat plate-glass that serves as the diagonal, a series of abrasive powders ranging from coarse to fine, a supply of optical rouge for polishing and a quantity of pine pitch.

As Russell Porter explained in 1925, "In the reflecting telescope, *the mirror's the thing*. No matter how elaborate and accurate the rest of the instrument, if it has a poor mirror, it is hopeless." Fortunately it is all but impossible to make a really poor mirror if one follows a few simple directions with reasonable care. The idea is to grind one face of the six-inch mirror-blank to a shallow curve about a 16th of an inch deep, polish it to a concave spherical surface and then, by additional polishing, deepen it increasingly toward the center so that the spherical curve becomes a paraboloid. The spherical curve is formed by placing the mirror blank on the tool, with wet abrasive between the two, and simply grinding the mirror over the tool in straight back-and-forth strokes. Nature comes to the aid of the mirror-maker in achieving the desired sphere, because glass grinds fastest at the points of greatest pressure between the two disks. During a portion of each stroke the mirror overhangs the tool; maximum pressure develops in the central portion of the mirror, where it is supported by the edge of the tool. Hence the center of the mirror and edge of the tool grind fastest, the mirror becoming concave and the tool convex. As grinding proceeds, the worker periodically turns the tool slightly in one direction and the mirror in the other. In consequence the concavity assumes the form of a perfect sphere because only mating spherical curves remain everywhere in contact when moved over each other in every possible direction. Any departure from a true sphere is quickly and automatically ground away because abnormal pressure develops at the high point and accelerates local abrasion.

The grinding can be performed in any convenient location that is free of dust and close to a supply of water. The operation tends to become somewhat messy, so a reasonably clean basement or garage is preferable to a kitchen or other household room.

A support for the tool is made first. This may consist of a disk of wood

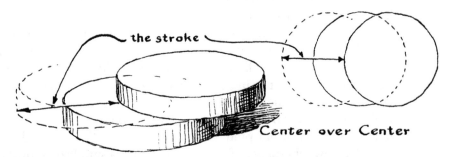

Figure 1.2 Details of the stroke used in grinding the objective mirror of the telescope

roughly half an inch thick fastened to the center of a square of the same material about a foot on a side. The diameter of the wooden disk should be about half an inch smaller than that of the tool. All surfaces of this fixture, except the exposed face of the wooden disk, should receive two coats of shellac. The glass tool is then cemented symmetrically to the unfinished face of the wooden disk by means of pitch. Melt a small quantity of pitch in any handy vessel. Warm the tool for five minutes in reasonably hot water, then dry it and rub one face lightly with a tuft of cotton saturated with turpentine. Now pour a tablespoon of melted pitch on the unfinished face of the wooden disk and press the tool against it so that pitch squeezes out all around the joint. After the tool and supporting fixture cool, they are a unit that can be removed from the bench conveniently for cleaning, which is frequently needed. Some workers prefer to attach the wooden disk to a large circular base. The base is then secured to the bench between three wooden cleats spaced 120 degrees apart. This arrangement permits the base to be rotated conveniently.

The tool assembly is now fastened on the corner of a sturdy bench or other working support, and a teaspoon of the coarsest abrasive is sprinkled evenly over the surface of the glass. A small salt-shaker makes a convenient dispenser for abrasives. The starting abrasive is usually No. 80 Carborundum, the grains of which are about the size of those of granulated sugar. A teaspoon of water is added to the abrasive at the center of the tool and the mirror lowered gently on the tool. The mirror is grasped at the edges with both hands; pressure is applied by the palms. It is pushed away from the worker by the base of the thumbs and pulled forward by the fingertips. The length of the grinding strokes should be half the diameter of the mirror. In the case of a six-inch mirror the strokes are three inches long—a maximum excursion of an inch and a half each side of the center. The motion should be smooth and straight, center over center, as depicted in the drawing above. Simultaneously a slight turn is imparted to the mirror during each stroke to complete a full revolution in about 30 strokes. The tool should also be turned slightly in the opposite direction every 10 or 12 strokes. Learn to judge the length of the stroke. Do *not* limit it by means of a mechanical stop. Beginners will tend to overshoot and undershoot the prescribed distance somewhat, but these errors average out.

Fresh Carborundum cuts effectively, and the grinding is accompanied by a characteristic gritty sound. Initially the work has a smooth, well-lubricated feel. After a few minutes the gritty sound tends to soften and the work has a gummy feel. Stop at this point, add another teaspoon of water and resume grinding until the work again feels gummy. Both the mirror and tool are removed from the bench and washed free of "mud," the mixture of pulverized glass and powdered abrasive that results from grinding. This marks the end of the first "wet." Fresh Carborundum is now applied, and the procedure is continued for three additional wets. The stroke is then shortened to a third of the diameter of the mirror (two inches in the case of a six-inch mirror) for two more wets. The mirror should now show a uniformly ground surface to the edge of the disk in every direction. If not, continue grinding until this is achieved.

The ground surface now has the form of a shallow curve and must be tested for focal length. This is easily accomplished on a sunny day. The test equipment consists of a square of light-colored cardboard about a foot across which serves as a screen on which the image of the sun is projected, and a supply of water to wet the roughly ground surface of the mirror and thus improve its effectiveness as a reflector. Stand the cardboard on edge at a height of about six feet so that one side faces the sun squarely; then take a position on the shady side about 10 feet from the screen. Dip the mirror in the water and, with the ground surface facing the sun, reflect sunlight onto the screen. The image will appear as a fuzzy disk of light, doubtless somewhat smaller than the diameter of the mirror. The size of the image will change as the mirror is moved toward or away from the screen. Find the distance at which it is minimum. This is the approximate focal length of the mirror. At this stage of grinding, the focal length will doubtless be of the order of 15 feet. The object is to shorten it to six feet by additional grinding. Wash the tool, apply fresh abrasive, grind for five minutes and repeat the test. It is advisable to make a chart on which the focal length is recorded after each spell of grinding. The chart aids in judging progress toward the goal of six feet. When the desired focal length is attained, thoroughly scrub the mirror, tool, bench, utensils and all other objects likely to be contaminated with No. 80 abrasive. Grinding is then continued with successively finer grades of abrasive. The same stroke is used: two inches in length and center-over-center. Usually the second grade is No. 180, which has the texture of finely powdered sand. The grinding technique is precisely the same for all subsequent grades of abrasive; each stage of grinding is continued until all pits made in the glass by the preceding grade have been removed. Usually six wets with each grade is adequate. On the average each wet will require about 15 minutes of grinding. Examine the ground surface by means of a magnifying glass after the sixth wet. If any pits larger than average are found, continue grinding for another wet or two and examine again. Persist until all pits larger than average disappear. There is one exception to this procedure. Sometimes a stray grain of No. 80 or one of the intermediate grades will find its way into work that has reached the terminal stages of fine grinding. A scratch or groove will appear that is so deep that it cannot be removed by a reasonable amount of fine grinding. The only solution is to return to the offending grade and repeat all the intermediate work. Gloves are notorious grit-catchers. Never wear them when grinding. Try to prevent clothing from coming into contact with loose grit. Abrasives supplied with representative kits include Nos. 80, 180, 220, 280, 400, 600, FFF and rouge.

Scratches can also be made by lumps that form in all grades of fine abrasive. The lumps plow grooves in the glass just as though they were solid particles. They can be dispersed by a sedimentation procedure that improves the abrasive in another respect. All grades of abrasive contain powdered grit: particles much smaller than those of the maximum size. When the powder becomes wet, it acts like mud in that it retards cutting action. By removing the powder the time required for the final stages of grinding can be cut in half.

Abrasives are graded by number, ranging from 80 (particles about the size of granulated sugar) to 600 (microscopic particles). The coarser grades do not clump and rarely cause scratches. The difficulty appears with grade 320 and smaller. To purify abrasives you will need a few jars of clear glass ranging in size from a quart to a gallon, small jars with caps to hold the purified abrasive, four feet of rubber hose a quarter of an inch in diameter and a quart of water glass (sodium silicate).

Put clean water, together with about two ounces of water glass, in a gallon jar until the level is an inch below the top. The water glass serves as a deflocculating agent: it disperses lumps that remain solid in water alone. One ounce of abrasive is thoroughly mixed with the solution and left to settle for 30 minutes. Siphon all but two inches of the fluid into a clean container. Label the container 600-1 and put it aside.

Refill the settling jar with water containing one ounce of water glass, thoroughly mix the remaining grit and again let it settle for 30 minutes. All but an inch of the fluid is then siphoned into a clear glass container and labeled 600-2. Thereafter, repeat the procedure, progressively reducing the intervals of settling to 15, eight and three minutes. The stored containers are labeled 600-3, 600-4 and 600-5 respectively.

Finally, shake up the settled dregs and pour them into a smaller jar. This material settles quickly. A sharp line appears at the boundary between the clear fluid and the suspended grit. When the upper third of the fluid clears, carefully pour all but a third of the remainder into a clean jar. When this material settles, pour off and discard the clear fluid. Then refill the jar that contains the dregs and repeat the procedure three times. The collected material is labeled 600-6. To the remaining dregs add one ounce of the 600 grit as it comes from the manufacturer, process it by the same procedure and similarly treat the remaining stock. After several days, when the grit in all six labeled containers has settled, carefully siphon off the clear fluid and dry the abrasives for use.

(continued)

What about the accumulated dregs? To them add one ounce of 500 grit, proceed as described and then switch to 400, followed by 320. Do not process the coarser grades.

Purified abrasive easily cuts twice as fast as untreated material. During the final grinding stage, when 600-6 grit is followed by 600-5, -4, -3, -2 and -1, the glass emerges unscratched and with a semipolished surface.

Ed.

The beginner is urged to purchase an extra mirror-blank. The object is to make two mirrors simultaneously, select one for immediate use and reserve the second for subsequent refinement. Those following this suggestion should grind the mirrors alternately. Complete a wet of a given grade on the first mirror and proceed with the same wet on the second. After all grinding is completed, the mirrors are polished independently.

The operations of grinding and polishing glass are similar in that both require the use of a material which is harder than glass. In grinding, the abrasive material is used between a pair of hard surfaces, either two pieces of glass or glass and cast iron. In rolling between the surfaces under pressure the hard particles erode the glass by causing tiny conchoidal fractures in its surface. Glass can be polished with the same hard particles merely by replacing the hard tool with an appropriately soft and yielding one. The abrasive particles do not roll. Held firmly by a yielding medium, their protruding edges may act like the blade of a plane.

Most amateurs use a polishing tool, or "lap," of pine pitch divided into a pattern of facets and charged with rouge. To make the facets, pitch is first cast into strips about an inch wide, a quarter of an inch thick and eight or 10 inches long. An adequate mold of wood (lined with moist paper to prevent the strips of pitch from sticking) is depicted on page 11. Melt the pitch over a hot plate, not over a direct flame. Do not overheat the pitch; it burns easily. The fumes (largely vaporized turpentine) are highly combustible, so prevent direct flame from reaching the open part of the container.

The strips of cool pitch are cut into square facets by means of a hot knife and stuck to the ground surface of the tool in a checkerboard pattern as shown. Begin by locating one facet somewhat off-center in the middle of the tool, and work outward. Adhesion is improved by first warming the tool, smearing it with a film of turpentine and warming the face of each square of pitch before placing it in contact with the glass. The pitch facets should be beveled, which can be accomplished in part by cutting the edges of the wooden divider-strips of the mold at an angle. This also facilitates the removal of the strips from the mold. Pitch yields under pressure, so

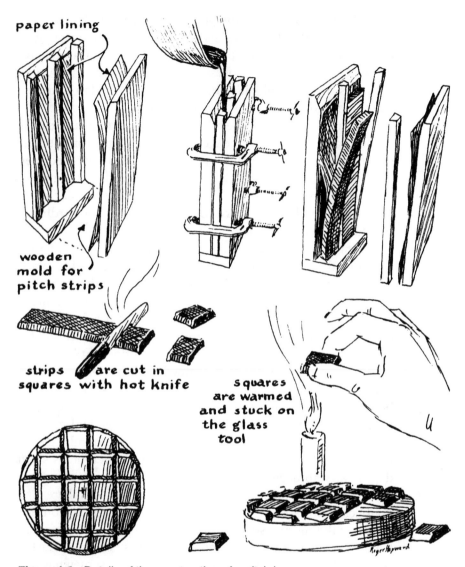

paper lining

wooden
mold for
pitch strips

strips are cut in
squares with hot knife

squares
are warmed
and stuck on
the glass
tool

Figure 1.3 Details of the construction of a pitch lap

unless the facets are beveled the space between adjacent facets soon closes
during the polishing operation.

Trim all boundary facets flush with the edge of the tool by means of
the hot knife. Then invert the completed lap in a pan of warm water for 10
minutes. While the pitch is warming, place a heaping tablespoon of rouge
in a clean wide-mouthed jar fitted with a screw cap, and add enough water
to form a creamy mixture. Remove the lap from the pan, blot it dry and,
with a quarter-inch brush of the kind used with water colors, paint the

pitch facets with rouge. Now place the mirror gently and squarely on the lap and apply about five pounds of evenly distributed weight to the mirror for half an hour until the pitch facets yield enough to conform with the curve of the glass. This process is called cold-pressing. At the end of the cold-pressing interval slide the mirror from the lap and bevel the edge facets to remove any bulges that have formed.

The mirror must now be fitted with a shield to insulate it from the heat of the worker's hands. In the case of a six-inch mirror cut a disk of corrugated cardboard eight inches in diameter and notch its edge every inch or so to a depth of one inch. Center the cardboard on the unground side of the mirror, press the notched edges down along the side of the glass and secure them with several turns of adhesive tape. The cardboard form now resembles the lid of a wide-mouthed jar.

Paint the facets with fresh rouge, add half a teaspoon of water to the center of the lap and, with the heat-insulating shield in place, lower the mirror gently onto the lap. Polishing proceeds with strokes identical with those used in grinding; they are two inches long and center-over-center. When the work develops a heavy feel, stop, add half a teaspoon of water and resume. Continue polishing for 20 minutes. Then cold-press for 10 minutes. Proceed with this alternating routine until no pits can be detected when the surface is examined with a high-powered magnifying glass. If the fine grinding has been performed as directed, the mirror can be brought to full polish in three hours or less. When work must be suspended for some hours, coat the lap with rouge and cold-press without added weight. It is well to brace the mirror around the edges when it remains on the lap for some hours, because pitch flows slowly and may deposit an unbraced mirror on the floor.

The shape of the mirror is now close to a perfect sphere. The center will doubtless have a somewhat longer radius than the region near the edge. Precisely the reverse situation is desired: a curve whose radius increases from the center outward. A minute thickness of glass must therefore be removed from the center of the mirror and a somewhat lesser amount removed toward the edge. The mirror is put back on the lap and, with a fresh charge of rouge, polished by a modified stroke. The length of the stroke is not altered, but the mirror is now made to follow a zigzag course laterally across the lap at right angles to the worker. The first stroke follows the conventional center-over-center course away from the worker but on the return stroke and subsequent strokes it is pushed about an inch to the right side. It is then gradually worked back across the center until it overhangs the lap on the left hand side by an inch. This operation is repeated over and over for 15 minutes. Simultaneously the mirror is rotated slightly

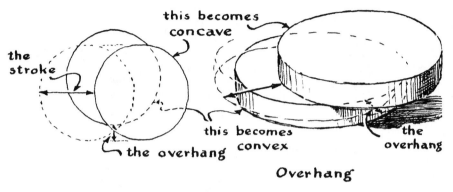

Figure 1.4 Details of the stroke used to parabolize the objective mirror

in one direction during each stroke and the tool is periodically rotated in the other direction to distribute the abrasive action uniformly.

After a thorough cleaning the mirror is ready for silvering. Amateurs formerly coated their mirrors at home. But silver is difficult to apply and tarnishes quickly. Most reflecting telescopes are now aluminized. The mirror is placed in a highly evacuated chamber and bombarded with vaporized aluminum. On being exposed to air the metal acquires a transparent and durable film of oxide. The beginner is urged to have both the objective mirror and the diagonal coated in this way by a company that specializes in this kind of work.

The Foucault test for determining the shape of a concave mirror, capable of accuracy to a millionth of an inch, is the essence of simplicity. You make a pinhole in a tin can, put a flashlight bulb inside and shine the rays from the pinhole (a synthetic "star") on the mirror. If the mirror has the figure of a true sphere, the reflected rays converge to form an image of the pinhole. When the mirror is viewed from a point just behind this image, it appears evenly illuminated and flat, like the disk of the full moon. And if you pass a knife-edge through the center of the image, the mirror should darken uniformly.

That is the way the test is *supposed* to work. In practice it is much more interesting—or exasperating, depending upon your temperament. The slightest departure of the mirror from a true sphere—or an equivalent change in its position or an abrupt variation in the density of the surrounding air—destroys the apparent flatness of the disk. With appropriate modifications of the light source, you can take advantage of this sensitive property and use the apparatus for photographing rifle bullets in flight complete with the shock waves. Similarly, you can photograph sound waves, convection currents, streamlines around airfoils and so on. The

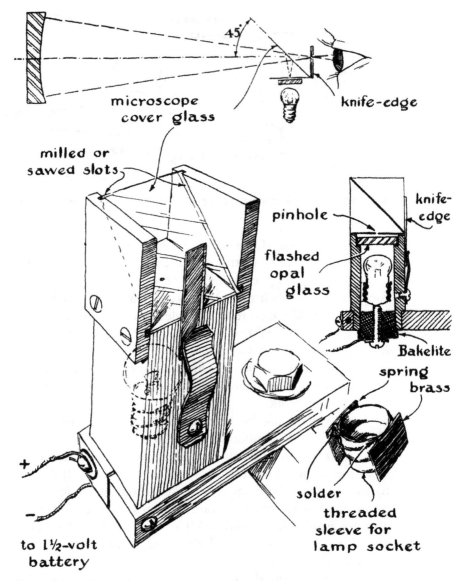

45°

microscope
cover glass

knife-edge

milled or
sawed slots

pinhole

flashed
opal
glass

knife-
edge

Bakelite

spring
brass

+

−

to 1½-volt
battery

solder
threaded
sleeve for
lamp socket

Figure 1.5 A Foucault-test rig for short-focus mirrors

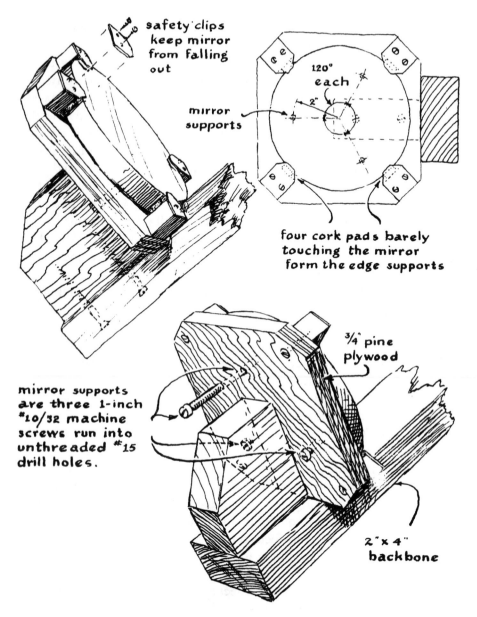

safety clips keep mirror from falling out

mirror supports

120° each

2"

four cork pads barely touching the mirror form the edge supports

mirror supports are three 1-inch #10/32 machine screws run into unthreaded #15 drill holes.

¾" pine plywood

2" x 4" backbone

Details of wooden cell for the objective mirror

Figure 1.6 Details of wooden cell for the objective mirror

setup can even be adapted as an ultrasensitive seismometer, which will pick up the vibrations of traffic miles away.

Amateur telescope makers have a lot of fun doing experiments like these. But primarily they use the test as the French physicist Jean Foucault intended it: for determining when the mirror has been polished to the figure of a parabola, the shape required for a good reflecting telescope. When the parabola is examined at the knife-edge, it presents a pattern shaped somewhat like a doughnut instead of a flat disk. As the ratio of the focal length to the diameter of the mirror is increased, the distinction between the two patterns tends to disappear. The doughnut becomes flatter with increasing focal length. At about *f*/15 the curve of the parabola coincides with that of the sphere for all practical considerations, and the Foucault pattern for both appears flat. Below *f*/5 the "doughnut" develops such pronounced shadows that interpretation becomes difficult and the test loses its usefulness.

Many amateurs have dreamed up schemes for making the Foucault apparatus less finicky. The trouble stems from the fact that, as conceived by Foucault, the pinhole must be situated somewhat off the optical axis of the mirror, so that the reflected image will form on the opposite side, where there is space for the amateur's eyeball. The strongly shadowed doughnut of short-focus mirrors results from this angular illumination.

Carl Bergmark, an amateur telescope maker of San Francisco, submits a solution he has devised for the problem:

"One of the defects of the Foucault test when working with short-focus mirrors is the error introduced by the lateral distance between the pinhole and the knife-edge. I believe that my test rig eliminates this shortcoming. The light from the pinhole is reflected into the axis by a microscope cover glass instead of being directed straight toward the mirror. This arrangement causes some loss of illumination, but the comparatively great light-gathering power of the short-focus mirror compensates for it. Rays reflected from the mirror are transmitted to the eye through the cover glass, the knife-edge being inserted at the point of convergence between them. Although the surfaces of the cover glass reflect a double image of the pinhole, the glass is so thin that the two images may be regarded as a point. I am lucky in having access to a nine-inch metal lathe, the compound rest of which serves as a handy carriage for the gadget. While under test the mirror is supported on a wall bracket behind the lathe."

After making the drawings to illustrate Bergmark's rig, Roger Hayward, who is an old hand at devising optical tricks, observed: "I have a feeling that Bergmark's scheme of using a microscope cover glass for an image divider is not altogether new, but I cannot remember having seen it applied in just this way. The secret of its success lies in the fact that, bad as cover glasses are optically, a very small area of one is always good enough. Incidentally, the formula for computing the difference between the position of the knife-edge at which the center of the mirror appears to darken uniformly and that at which any radial zone of a parabola similarly darkens is computed by the equation: $D = r^2/2R$, where r is the radius of the zone, R the radius of the mirror's curvature and D the difference in position through which the test rig must be moved to achieve the desired darkening."

The mounting may be constructed while the mirrors are being coated. In designing the mounting never permit appearance to compromise sturdiness. This telescope will have a maximum magnifying power of about 250 diameters and any jiggle arising in the mounting will be magnified proportionately. The objective mirror is supported in a wooden cell fitted with screw adjustments, as shown in detail on page 15. Fine-grinding and polishing will have reduced the focal length to about five feet. The center of the diagonal mirror is spaced approximately six inches from the focal point of the objective (about 4.5 feet from the objective), thus bending a six-inch cone of rays into the eyepiece.

After assembly the optical elements must be aligned. Remove the eyepiece, look through the tube in which it slides and adjust the diagonal mirror until the objective mirror is centered in the field of view. Then adjust the tilt of the objective mirror until the reflected image of the diagonal mirror is centered. Replace the eyepiece in its tube and you are in business.

2 HOW TO GRIND, POLISH AND TEST AN ALUMINUM TELESCOPE MIRROR

Conducted by C. L. Stong, November 1963

Amateurs who make small telescopes tend to overlook the virtues of metal mirrors. This is not to suggest that better mirrors can be made of metal than of glass, even in the case of small instruments; when the advantages of the two are compared, glass usually emerges as the preferred material. Glass takes a good polish, its reflecting film of metal can be replaced easily and it retains its shape except during periods of changing temperature. Glass held a decisive advantage over metal in the days when mirrors were silvered, since tarnished silver can be removed from glass and replaced inexpensively at home in a single evening. Now aluminum has replaced silver as the reflecting surface. Aluminum acquires a protective film of oxide and with reasonable care the surface retains its brightness for years. But the aluminum film is applied to glass by a process of evaporation in a vacuum that requires an apparatus beyond the reach of most amateurs, who must send their mirrors to a commercial establishment for resurfacing. A small solid aluminum mirror, on the other hand, can be repolished at home in an evening. Preserving the shape of the reflecting surface during repolishing can be troublesome, however, and until the novice gets the knack repolishing may take the instrument out of service almost as long as the aluminizing procedure would. Once it has been repolished, the surface of the solid metal acquires the oxide coating, stays bright for a long time and retains its optical figure during changes of temperature that would put a glass mirror out of business.

W. C. Peterson, an amateur telescope maker of Pittsburgh, Pa., made his first aluminum mirror in 1943 and has not touched glass since he switched to metal. "In brief," he wrote, "my process involves two disks of metal, one for the mirror and one that serves as a tool. One surface of the mirror blank is made concave and one surface of the tool convex so that the pair mate like a shallow ball and socket. The metal can be worked easily with a scraper and file if the experimenter does not have access to a lathe. The roughed-out blanks are ground together with successively finer grades of abrasive until their surfaces mate. Then the concave member of the pair is given a prepolish with pumice and finished like a mirror with rouge on a lap made of hard pitch.

"I have made excellent mirrors of stainless steel but advise the novice to begin with aluminum. Any of the hard bright aluminum alloys work well. They come in the form of bar stock and odd lengths can be procured occasionally from dealers in nonferrous metals. I recommend for an introductory exercise a pair of blanks in the form of disks three inches in diameter and ½ inch thick. The thickness must be at least a twelfth of the diameter so that the blank will not flex during the grinding operation and distort the desired curvature, but certainly it need not be thicker than an eighth of the diameter.

"I begin by drilling a carefully centered hole about ⅟₁₆ inch in diameter and ¼ inch deep in one side of each blank as a reference center. I also make four disks of hardwood of the same diameter, about ¾ inch thick, and shellac them to seal the wood against moisture. Their use will become apparent.

"The next requirement is a pair of sheet-metal templates to serve as guides for roughing the blanks to the desired curvature. The depth of the curve determines the ratio of the diameter of

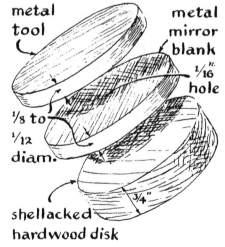

Figure 2.1 Components of the metal mirror

the mirror to its focal length, the *f* number. In my experience—and most amateur telescope makers will agree—the relative aperture should not be more than *f*/8 or less than *f*/10, with *f*/9 as a fine compromise. In the case of a three-inch mirror, a focal length of 27 inches would represent a good choice. It is not always possible for the beginner to grind a curve that hits the specified focal length on the nose, but by aiming for 27 inches one can usually achieve a curve that ranges between 24 and 30 inches and is therefore within the accepted limits. The radius of the curvature is equal to twice the focal length. The radius of an acceptable three-inch mirror should therefore fall somewhere between 48 and 60 inches, with 54 inches the best length. I improvised a compass with which to scribe this radius: a six-foot stick with a screw at one end and an ice pick at the other. With the end of the stick screwed to the floor, the ice pick is inserted through a hole in the other end 54 inches from the screw. A three-inch-square sheet of zinc or hard brass is tacked lightly to the floor and the scriber is guided across the middle of the sheet to cut a deep groove completely across the metal. Then the sheet is flexed until it breaks along the arc, and the edges are dressed lightly with a file. The halves serve as the templates, one convex and one concave.

"To make a tool for roughing out the curve of the mirror, grind the end of a flat file to the shape of a thumbnail for use as a scraper [*see illustration below*] and wrap the body of the file with electrical tape for a handle. With this tool scrape one side of the blank selected for the mirror until its curvature fits the convex template. This may sound like a job for a lathe, but the work can be done about as easily by hand. Aluminum is soft and only a small amount of metal must be removed—less than half the thickness of a dime. Then use a file to shape the other blank convex to fit the concave template. Do not strive for precision, but try to avoid deep gouges.

"The unscraped side of each blank is now cemented to one of the disks of hardwood. I use common roofing tar as cement—the kind that comes in lumps—and flow it onto the work by heating it with an electric soldering iron. A thin layer of tar is applied to the metal and the wood and the disks are simply pressed together. Seal any crack that develops between the disks by applying the hot iron. Before cementing the mirror to its wood backing I drill a hole through the wood disk large enough for a No. 6 machine screw and attach a nut to the inner

Figure 2.2 Shaping mirror with homemade scraper

face of a metal plate that is then recessed over the hole. This provides a convenient fixture on which to mount the mirror in the telescope.

"Next I make a shallow wooden tray about six inches square and one inch deep, with three cleats screwed to the bottom 120 degrees apart and spaced to make a snug fit with the convex tool blank. The tray should be attached rigidly to a firm bench that is about waist-high. Mount the tool in the tray, apply about a quarter of a teaspoon of carborundum grit to the tool and wet it with an equal amount of water. Invert the mirror over the tool and grind by pushing the mirror back and forth. The length of the strokes should be adjusted so that the mirror overhangs the tool about ½ to ¾ inch at the end of each stroke. The center of the mirror should pass directly over the center of the tool. Only two grades of carborundum are used: 220 mesh and 320 mesh. The work will require less than a quarter of a pound of each grade in the case of a three-inch mirror.

"The length of the stroke is not critical, but short strokes make the curvature shallow and long ones deepen it. Rotate the mirror slowly while stroking and work it around the tool to distribute the grinding uniformly. Add water to the carborundum from time to time and replace the grit as it turns to mud and becomes ineffective. There is no hard and fast rule for adding water and replacing grit, but you develop a feel for the procedure rather quickly. Grit makes a grinding sound when it is working well, and the mirror slides over the tool with little effort. Spent grit should be wiped from the metal with a rag. (Do not flush it down the drain because it will probably clog the plumbing.)

"When the surfaces of both blanks are fully ground, flush the mirror with clean water and while it is still wet reflect an image of the sun against a wall or a screen of cardboard. Move the mirror toward and away from the screen until the sun's image is sharpest (smallest); the distance between the mirror and the screen should be between 24 and 30 inches. If it is not, check the accuracy of the templates and if necessary make up a new set and start again from the beginning. When you are satisfied that both blanks have been accurately ground with 220-mesh grit, switch to 320 and continue until all evidence of the coarser abrasive disappears.

"The next procedure may sound strange to glassworkers, although it is not new. A polishing lap is prepared of hard pitch—one that would selectively deepen the curve in the center of a glass mirror and result in what experienced telescope makers refer to as 'the fatal hyperbola' or 'a turned-down edge.' Although an extremely hard lap is rarely used for glass, it works like magic on aluminum and accounts for the ease with which beginners can make metal mirrors. One of the remaining wood disks now comes into play. Chunk tar of the roofing variety is first melted with the soldering iron and flowed over the wood to a depth of ⅛ inch. It will have little tendency to overflow. Then about a third to half as much lump rosin is melted, flowed into the tar and thoroughly blended with it. (Powdered rosin will not mix with tar. If rosin is available only in powdered form, melt it and after the batch cools break it into lumps.)

"Paint the surface of the warm pitch with polishing rouge that has been mixed with water to the consistency of heavy cream, place the concave face of the mirror squarely over the painted surface and swirl the mirror until the curve of the pitch conforms with that of the mirror. 'Press' the assembly by allowing it to stand and cool to room temperature. If pockets or bubbles are found in the pitch when the mirror is subsequently removed from the lap, use some of the runoff for patching the holes. Flow in just enough of the tar-rosin mixture to fill the holes. Then paint the patches with rouge and press the entire lap with the mirror as before. To test the pitch for hardness, make a firm cut across the lap with a wet knife; the pitch should splinter and make a crackling sound. To soften add tar, to harden add rosin. The edge of the lap is then trimmed with the wet knife.

"The prepared lap is now charged with 320-mesh carborundum (not rouge!) and stroked with the mirror as during rough grinding. I do not favor any form of circular stroke, but one must continuously rotate the mirror and more or less work around the tool in all possible orientations to preserve the element of randomness. Now to the crux of the procedure: The worker must examine the pitch lap every two or three minutes, re-press it if necessary and occasionally remake it when the tar and rosin become thin at the edge. I re-press for 10 to 15 minutes after every 10 or 15 minutes of polishing. If the room is cooler than about 70 degrees Fahrenheit, I warm the lap under a hot-water faucet before pressing.

"In an astonishingly short time the mirror takes on some polish and forms a clear image of the sun even when dry. The focal length can now be confirmed more accurately with a Foucault-test apparatus. In principle this test consists of viewing the reflected image of a pinhole source of light—an artificial star. Both the pinhole and the observer's eye must be at a distance from the mirror that is equal to just twice the focal length of the mirror. Of course, when the pinhole is precisely twice the focal distance from the mirror and squarely on the optical axis of the mirror, the focal point will fall on the pinhole itself. But if the pinhole is shifted slightly to the right of the optical axis, its focused image will shift the same distance

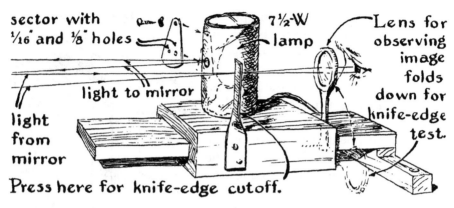

sector with
¹⁄₁₆" and ¹⁄₈" holes

7½-W
lamp

Lens for
observing
image
folds
down for
knife-edge
test.

light to mirror

light
from
mirror

Press here for knife-edge cutoff.

Figure 2.3 Details of Foucault-test apparatus

to the left, where it can be observed. The image can be found by exploring the general area with the aid of a ground glass until a spot of light appears on the glass. When the eye is moved to a position a foot or so directly behind the image and the ground glass is taken away, a minute 'star' will be seen hanging in mid-air. This is the image of the pinhole. If the eye is now brought close to the image, the face of a fully polished mirror of spherical figure will be seen as a glowing disk that resembles the full moon, because light from the pinhole strikes every part of the mirror and is reflected equally from every part into the eye. The image of the pinhole is real, and it can be examined with a magnifying glass.

"The complete Foucault-test apparatus includes a bracket for supporting the mirror, a lamp house, a movable sector containing the pinholes, a magnifying glass and a knife edge—all mounted on a base fixture that moves on a rail, as shown in the accompanying illustration [*above*]. When the image of a large pinhole is examined during the early stages of polishing, the edge will appear to be irregular and the face of the metal will have a grainy texture caused by myriad pits. Often the surface will resemble the rough skin of a tangerine. As polishing continues, the image will gradually become disk-shaped and the colors will simultaneously change from chocolate through orange and yellow to brilliant white, even when the smallest pinhole of the apparatus is used.

"The polish rarely progresses evenly to the edge of the mirror, even though the tool is kept true. To correct the tendency of the center to polish first, I cut a lopsided, long-armed star with four or five points from the middle of the lap and extend the tips of the star to within half an inch of the edge. This has the effect of accelerating the action of the abrasive toward the edge. In spite of directions given by many books for polishing glass, never remove pitch from the edge portions of the lap when making a mirror of metal. If the edge fails to polish after treatment with the star lap, the worker has missed the boat somewhere along the line and must return the mirror to the tool and try again.

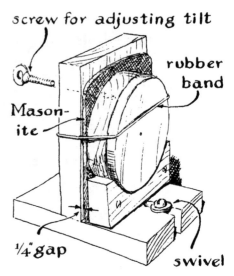

Figure 2.4 Test bracket for supporting mirror

"When the pits have been reduced substantially, switch from 320-mesh carborundum to pumice. I find that conventional kitchen cleansers such as Ajax work splendidly. To make the change simply remelt the used lap, paint it with rouge and press. After the lap cools charge it with pumice and resume polishing. (The presence of embedded carborundum in the pitch does no harm.) After a few spells of polishing with pumice finish each period by adding rouge to the lap. After a while examine the image of the pinhole with the magnifying glass. At first it will appear as a relatively large, fuzzy patch of light, but as the polishing continues details will stand out with increasing clarity; bits of lint and the rough edges of the metal will be seen highly magnified. In effect, the mirror has started to function as the objective of a telescope. When the pinhole can be seen in sharp detail and all evidence of pits has vanished, make a new lap and switch from pumice to rouge.

"The new lap for rouge must be made of clean tar and rosin, uncontaminated by carborundum or pumice. Use the remaining hardwood disk for this rouge lap and clean the working area thoroughly of all carborundum and pumice. Combine the tar and rosin in the same proportions as before. During succeeding spells of polishing, the rouge should change color promptly if all is going well. If the switch from pumice to rouge is made before all pits have been removed, the rouge may remain dark and the mirror will not take a brilliant polish. In that case go back to polishing with pumice. Laps sometimes misbehave, however, even when all pits have been removed. The mirror may tend to stick or to pick up pitch. This difficulty can usually be cured simply by adding a drop of ordinary mucilage to the wet rouge. If the mucilage fails to work, make a new lap. At this stage of polishing, incidentally, the mirror should never be removed from the lap for extended periods. Grooves about ⅛ inch in diameter should be cut through to the wood so that the pitch is divided into rectangles, each about an inch square. The pattern of rectangles should not be distributed symmetrically with respect to the center of the tool.

"After a brilliant polish is achieved from the center of the mirror to the edge, as judged by eye, begin to use the Foucault knife-edge test. Turn the magnifying glass to the side, align the image of the pinhole so that it almost grazes the 'knife' and, with the eye directly behind the image so that the mirror is seen as a full moon, press the knife into the light rays. When the knife is between the mirror and the image, the apparent shadow cast by its edge will move across the face of the mirror from right to left. When the blade is between the image and the eye, the shadow will cross in the opposite direction, from left to right. Manipulate the blade until it cuts the focal point of the rays. The mirror will then darken when the knife is moved and no shadow will cross the disk.

"If the curvature of the mirror is a perfect sphere, the surface will appear to be flat. If the knife blade is now moved very slightly ahead of the focal point, the surface will appear to be convex, like a ball, and if moved slightly behind, the surface will appear to be concave or cup-shaped. In the case of an *f*/9 mirror this is the desired test pattern; the beginner would be lucky indeed if it appears early during the polishing procedure. Usually a disk will be seen that has either a pronounced bulge or a depression in the middle. Such figures are corrected by altering the lap—removing pitch as required—or by changing the length of the polishing stroke, or both. Strokes that result in the mirror overhanging the lap by more than about half an inch tend to deepen the center, to correct humps or bulges. Those shorter than the normal half-inch overhang tend to bring up the center (or to depress the edges). Continue to polish until the whole surface of the mirror darkens uniformly without bulges or depressions when the knife cuts the rays from a pinhole ¹⁄₁₆ inch in diameter at the focal point. This completes the mirror.

"Reflecting telescopes of many types have been developed during the many years since Isaac Newton invented the instrument, but I prefer the simple version contrived by John Hadley, an English experimenter, in the early 18th century. The optical assembly of my version of this instrument is supported in alignment by a heavy tube of cardboard of the kind on which rugs are rolled. It is strong, easy to cut and thick enough to take wood screws, even at the ends. I always saturate the screw holes with shellac for extra strength.

"The mirror is mounted on a disk of plywood, large enough to cap the end of the tube, by means of a machine screw that engages the nut recessed in the wood block of the mirror. Three equally spaced wood screws fasten the assembly to the lower end of the tube. The holes for the screws are equipped with rubber grommets and the axis of the mirror is aligned with the axis of the tube by adjusting the screws.

"An oblong hole can now be cut in the side of the tube near the top for admitting the eyepiece assembly. This assembly includes a small front-surface mirror for deflecting the rays to a focus just beyond the outer edge of the oblong hole. The position of the center of the hole is determined by subtracting the radius of the tube from the focal length of the completed mirror. Both the small mirror and the lenses for a variety of eyepieces [*see illustration on page 27*] are available from the Edmund Scientific Company, 101 East Gloucester Pike, Barrington, NJ 08007 (*www.edmundscientific.com*).

"The base of the Hadley telescope resembles a three-legged milking stool 20 inches high. The top is 10 inches long and about half as wide. A single bolt in the center attaches a trunnion assembly that rotates in azimuth on the bolt. A pair of bolts extending outward from the middle of the tube constitute the elevation axis. They engage slots in the trunnion and attach to the tube through a pair of metal plates screwed to the cardboard. The tube is held in the trunnion slots by a pair of helical springs.

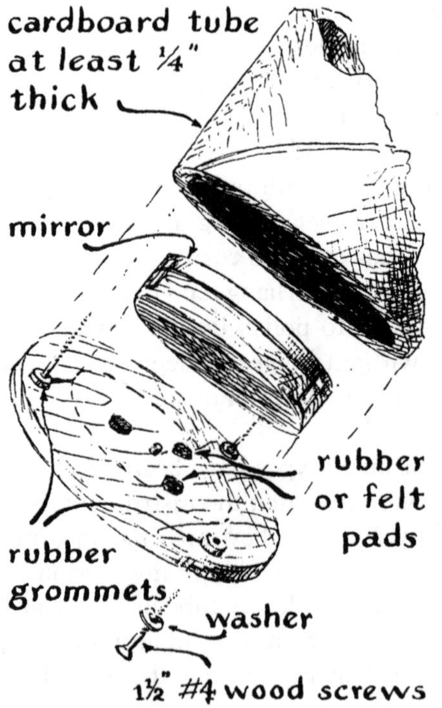

"Slow-motion drive in both azimuth and elevation is provided by a pair of miniature winches made of ¼-inch rods that fit the plastic knobs of an old radio set. An outrider on the base supports the azimuth winch and its cord, which swing a lever arm fastened to the trunnion assembly. The elevation winch is built into the end of the lever arm and its cord is fastened to the end of the tube. The bottom of the tube carries a small counterweight that keeps the elevation cord under tension. By modern standards the arrangement may seem somewhat primi-

cardboard tube at least ¼" thick

mirror

rubber or felt pads

rubber grommets

washer

1½" #4 wood screws

Figure 2.5 Details of mirror cell assembly

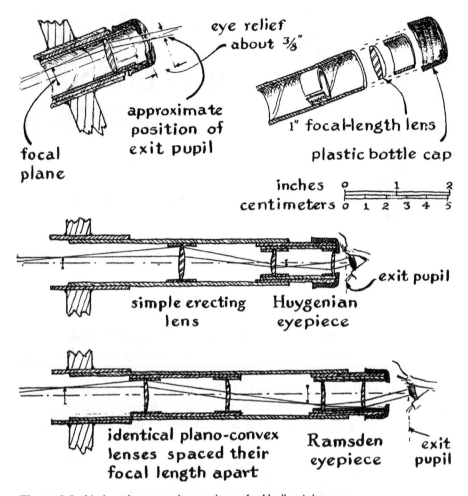

eye relief about 3/8"

focal plane

approximate position of exit pupil

1" focal-length lens

plastic bottle cap

inches

centimeters

simple erecting lens

Huygenian eyepiece

exit pupil

identical plano-convex lenses spaced their focal length apart

Ramsden eyepiece

exit pupil

Figure 2.6 Various homemade eyepieces for Hadley telescope

tive. I find that it has the offsetting virtues of low cost and simplicity. During the past three decades I have built many telescopes ranging in design from a replica of Newton's tiny instrument to the highly mechanized types so popular today. None is more convenient or pleasant to use, in my opinion, than Hadley's primitive design."

3 THE BEHAVIOR OF THE TELESCOPE-MAKER'S PITCH

Conducted by Albert G. Ingalls, October 1952

Every amateur astronomer who has polished a disk of glass on a pitch lap to make a telescope mirror knows that pitch slowly flows. Though pitch is a solid at ordinary temperatures, it is classed by science as a liquid with a viscosity many billions of times greater than that of water. The term viscosity has several meanings: to most people it suggests stickiness: to the physicist it means a resistance to flow arising from internal fluid friction.

It is a strange fact that the writers of instruction books on telescope making have not explained why a pitch lap does its work. They only tell how. Thus it has remained for a ringside observer to state it. The experimental physicist John Strong writes: "The viscous nature of the pitch prevents quick changes in the polishing surface so that the polishing pressures are greatest and the polishing action is greatest on areas of the work where the glass surface is relatively high."

After a single reading this sentence may seem to be no more than a simple statement of the obvious, yet it is an almost unique expression of a rather subtle process. Amateur telescope makers are often dissatisfied with the empirical; then they seek the underlying reasons for the phenomena of their art. The following discussion should not be confused with others that treat with the nature of the process that polishes glass; it has solely to do with the behavior of the pitch.

The sentence by Strong deals with the viscosity of pitch, but pitch is also elastic. Some optical workers would challenge this statement. Lively arguments have occurred between those who say, like John M. Pierce, "Pitch is certainly elastic, otherwise it would be no good for non-spherical surfaces such as paraboloids," and others who answer, "Liquid yes, and viscous yes, but elastic no—not so long as the indentation of my thumbnail remains in a lap." One advanced amateur optical worker comments: "The elastic recovery of pitch is new to me." When asked whether the same substance could be both viscous and elastic, a professional optical worker replied simply, "No." The question was submitted to Strong. "Is there a restoring force (elasticity) in pitch?" He answered, "Yes." He was also asked: "How can pitch be viscous and elastic too?" His answer was, "See the bouncing putty." True enough, silicone putty is viscous, elastic and visco-elastic. A ball of it will bounce, but when left on a table for a few hours the same ball will flow under the force of gravity into a pancake.

Possibly amateur optical workers, including the writer, should keep up to date with the sciences that touch upon their hobby. To attempt a remedy, the discipline of rheology was frontally attacked in the hope of finding an unequivocal answer about pitch. Rheology is the science of the deformation and flow of matter: gaseous, liquid and solid. The prefix "rheo" means flow, as in rheostat or diarrhea. Rheology deals with elasticity, viscosity, plastic flow, creep and other phenomena in liquids and resinous materials, of which pitch is an example, and in metals and other forms of matter, including crystals.

My investigation of the subject resembled a dog's attack on a porcupine. I did, however, find one simple demonstration that should convince anyone. Twist a length of pitch into a spiral. Release one end and the pitch untwists a little. The addition of a pointer may be needed to reveal this elastic effect. A curve was also found in E. N. da C. Andrade's *Viscosity and Plasticity* that describes the flow and recovery of "a pitch-like substance." Would it be safe to assume that this was also the curve for the actual pitch that is used by the optical worker? Even if so, would not the attempt of a non-rheologist to write an interpretation of it be brash, considering that rheology is slippery even for the rheologist? The following letter was written for the amateur telescope maker with pitch in his hair by W. H. Markwood, a vice-president of The Society of Rheology.

"As an interested reader of 'The Amateur Scientist' I am pleased at the opportunity to tell you about the rheological behavior of pine pitch during the polishing of optical surfaces. To begin with, materials of this sort are both elastic and viscous. However, one may not say that they have only one elasticity and/or one viscosity even at only one temperature. They are also influenced by how fast one makes them flow, that is, how hard they are pushed. Let us confine ourselves to the region between room temperature and the softening point (both wood and gum rosins 'melt' at about 180 degrees Fahrenheit). In this range, if a resinous material is pressed very gently it flows in the manner of a very thick oil, and stays put when the force is removed. It acts as though it were viscous only. On the other hand, if a great force is suddenly applied, it appears to react like an entirely elastic crystal. The applied force may bounce back or the 'crystal' may shatter.

"For intermediate forces and intermediate rates of force-application it seems to behave in three ways. There appear to be (1) a purely elastic component like a time-independent ideal spring, (2) an elastic component that will slowly recover after force removal and (3) a non-recoverable, viscous component. The middle one is sometimes called creep or visco-elasticity or elasto-viscosity or recoverable elasticity or retarded elasticity. The ideal spring part deforms instantly in direct proportion to the force applied, while the viscous part deforms, that is, flows, at a *rate* that is proportional to the force. However, the visco-elastic deformation is dependent not only on how much force but on how fast it is applied. This idea of 'parts' can be confusing. A thing is a thing and not three things. This 'thing' has what might be called a multi-behavioristic attitude.

"Let's examine the curve you traced from Andrade's *Viscosity and Plasticity*. The curve was intended to illustrate the above. If the ordinate represents deformation under a *constant* load, if the abscissa is time, and if the maximum on the plot is the time at which the load was suddenly removed, all the 'parts' described above are illustrated. At first a sudden, elastic jump occurs, deformation then slows down and becomes a combination of visco-elastic and viscous distortion for a while (curved part), then straightens out into 'pure' viscous flow. When the load is removed, an elastic rebound takes place that is equal to the initial jump; then a slow, time-dependent, retarded-elastic recovery appears; finally the material stops

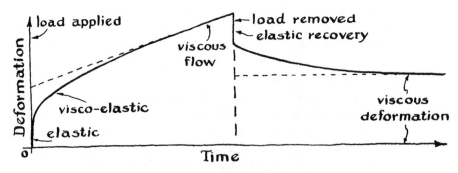

Figure 3.1 How pitch behaves when loaded and unloaded

recovering and it is seen that a 'set' has taken place that is permanent and represents the viscous deformation only.

"One can go further and visualize models to describe phenomenologically this 'multi-behavior.' Let the purely elastic part be represented by an ideal spring, the purely viscous part by an oil-filled dashpot, and the visco-elastic part by a spring whose action is damped by a dashpot connected rigidly parallel to it. If these are added together the situation shown in the enclosed drawing [*below*] obtains.

"Suppose one suspends this model vertically and hangs a weight on it. At the instant the weight is released the top spring stretches until the elastic modulus times the deformation of the spring just balances the weight. At this point (still zero time) the second spring begins to stretch, but its motion is damped by the dashpot, which first carries all the load, gradually transferring it to the spring until *its* elastic modulus times *its* strains also balances the effect of the weight, when its motion stops. However, the piston in the viscosity dashpot has also been moving and will continue to move linearly with time so long as the weight is applied. If the weight is now removed, the elasticity spring will snap back to its original position in zero time; the visco-elastic spring will try to do the same but will again be retarded by the 'oil flow' in its dashpot and will arrive back at its initial position in exactly the time it took to stretch. In the meantime the whole viscosity dashpot will go along for the ride; that is, its piston will not change position relative to its cup. This use of a model is a pretty crude but somewhat useful conception in a physical sense. It is often employed by the rheologist. Furthermore, the concept is easily handled mathematically.

"Looking again at the model, it is easy to visualize that if you pull on it fast enough and let it go fast enough only the elastic spring will stretch and recover. If you pull on it very, very slowly the viscosity dashpot will move before either spring can stretch without recovering in an infinitely short time as deformation proceeds—which brings us back to paragraph four.

"If the constant load we picked had been a different one, the behavior *pattern* would have been the same, but the relaxation time would have been different.

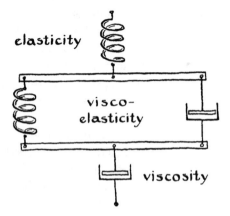

Figure 3.2 A model for pitch behavior

There is not just one but a whole spectrum of them, corresponding to spectra of elasticities and viscosities.

"I believe that if you'll warm some pitch, press on it hard, remove the load and watch, you'll see some small, slow recovery, indicating visco-elasticity. If you strike it rapidly and lightly, no change in shape will be seen. If you press on it very slowly and remove the load, no recovery will occur. Of course, its own mass will make it flow also; which can be seen if it is either warm enough or if you wait long enough."

Now that Markwood has assured us that pitch really is elastic as well as viscous, it is interesting to note that the great 19th-century English astronomer Sir John Herschel suspected the same thing from his optical shop experience. It is probable that he heard it from his famous telescope-making father William Herschel. John Herschel also believed that if pitch were not elastic, a mirror could not be parabolized on it. The following quotation from the younger Herschel's book *The Telescope*, published in 1861, concerning the polishing and parabolization of speculum-metal mirrors on a pitch lap, contains crystal-clear proof that both facts were known to him a century ago. For some reason they have not been restated in recent books on the art. Herschel wrote:

"When the metal is reduced by grinding to a perfectly true and even surface, free from the smallest perceptible scratch, it will be found reflective enough to afford an image of a star, or of a distant white object, sufficiently distinct to try whether its focal length is correct; and if it be so, the process of polishing may be commenced, the object of which is not merely to communicate a brilliantly reflective surface, but at the same time a truly parabolic form. If the material of the tool on which this operation is performed were perfectly hard and non-elastic, it is evident that this would be impracticable, since none but a spherical form could arise from any amount of friction on such a material once supposed spherical; and even if parabolic it could not communicate that form to a more yielding body worked upon it. . . . Happily, however, there exists a material which, with sufficient hardness to offer a considerable resistance to momentary pressure, is yet yielding enough to accommodate its form to that pressure when prolonged, and at the same time sufficiently elastic to recover it if quickly relieved; that substance is pitch, whose properties, in this respect, were at once taken advantage of by Newton, with that sagacity which distinguished all his proceedings, as the fitting material for a polisher."

Herschel, who so clearly understood the rheological behavior of pitch, was mistaken about Newton, who did not understand it. As is well known, Newton was unable to parabolize the two-inch spherical mirrors he made for the first reflecting telescopes in 1668 and 1671, and his description of

Most optical workers buy their pitch retail and cannot trace it through trade channels to its origin. Some use mineral pitch but many dislike it. Most of the pitch used in optical work is pine pitch. Asked about pine pitch Markwood replies: "Pitch is a general term applied to many natural and proprietary mixtures. There are two sources of such materials which, with other products, are commonly called naval stores. The first is from the living pine tree. When the tree is wounded by scoring it near the base, an oleoresin exudes which is a mixture of turpentine and rosin. These are separated by distilling out the turpentine.

"The second source is the stumps and heartwood of large virgin pine trees. By far the largest method of processing this wood is to grind it, extract it with a hydrocarbon solvent and distill out the oils to give a residue called wood rosin. Much of the rosin produced in this manner is refined by the use of selective solvents or adsorbents to give pale-colored rosin.

"Another method of processing pine-stump wood is to destructively distill it in retorts, using the technique employed for producing charcoal from hardwood. Pine-tar oil is volatilized from the wood. Most of the volatile material is removed from the product, leaving pine pitch. It softens at a lower temperature than rosin and contains less acidic material. Many of the commercial pitches sold under various names are obtained by this retort process. The compounded pitches are probably made either by modification of these materials or from rosin or related materials by plasticizing them with pine oils or other softeners to obtain the softening point desired."

Ed.

their polishing, in *Opticks,* shows why. He used a pitch lap "as thin as a groat," so thin that its elasticity was too insignificant to be effective; a groat was thinner than a well-worn dime. The probable explanation is that Herschel wrote without referring to *Opticks.*

4 THE DALL-KIRKHAM TELESCOPE

Conducted by Albert G. Ingalls, September 1951

In spite of a widespread impression, it is not easy to know the inventor and the date of invention of such things as the steamboat, the telegraph and the incandescent lamp. The same applies to the reflecting telescope. He who turns to history to find the solid ground of fact will soon find himself bogged in ooze. Was the inventor of the reflecting telescope the English astronomer John Hadley, who in 1723 made the first good modern reflector with a paraboloidal mirror? Or was it Sir Isaac Newton, who in 1668 built the first reflector with an eyepiece, though he knew no way to parabolize his spherical mirrors? Was it the Scottish mathematician James Gregory, who five years earlier proposed the type of reflector known today as the Gregorian, though for lack of manual dexterity he did not build one? Was it the French mathematician Marin Mersenne, often called Mersennus, who proposed the idea of a reflecting telescope to the French philosopher René Descartes in 1639, only to be told that it was fallacious? Was it the English mathematician Leonard Digges, who used concave mirrors before 1571, though probably without an eyepiece and as terrestrial telescopes? Was it the English philosopher Roger Bacon, who with Peter Peregrinus spent three years and the equivalent of $3,000 learning to make concave mirrors in approximately 1267? That Bacon even then understood spherical aberration is shown by his statement that the focal length is much less for rays from the outer zones of the mirror, and that it is half the radius of curvature. He left no record of the uses of the two mirrors that he made, but L. W. Taylor of Oberlin University tells us that a century later Peter of Trau recorded the tradition that with Bacon's mirrors "you could see what people were doing in any part of the world." He adds that the students at Oxford University spent so much time experimenting with these mirrors that the University authorities had them smashed. Perhaps their "experiments" were not limited to pure science.

It is thus that the invention of the reflecting telescope is blurred. Like most inventions it was a gradual process. The English mathematician Robert Smith, in his *Compleat System of Opticks,* published in 1738, refers to Gregory as "the first inventor" of the reflecting telescope; but Sir John Pringle in his *Discourse on the Invention and Improvement of the Reflecting Telescope,* delivered in 1777 before the Royal Society of London, designated Newton as "the main and effectual inventor." Pringle looked down his nose at the telescope which the French sculptor Guillaume Cassegrain revealed in 1672, describing it as merely "a disguised Gregorian never put into execution by its author"; but Louis Bell has pointed out in *Popular Astronomy* that "it is the irony of time that Cassegrain's form is the one that has survived in the great telescopes."

It may be surprising that Mersenne, Digges and others did not build directly upon the advances of their predecessors. Because of the lack of facilities for disseminating such information, they no doubt remained unaware of those advances. Galileo, who lived from 1564 to 1642 and built the first astronomical telescope, did not know that Leonardo da Vinci, who lived in Italy a century earlier, had designed machines for grinding concave mirrors; none of Leonardo's scientific writings were published until 1880, and they have not all been published yet.

In 1885, two centuries after Cassegrain's invention, a modification of the optics of that telescope was proposed in *English Mechanics* by an unknown contributor with the initials A.S.L. He proposed to substitute a simple sphere for the hyperboloidal secondary and to shape the paraboloidal primary so as to balance the aberrations of the secondary. This would call for an ellipsoid, sometimes loosely called an "undercorrected paraboloid." So far as is known nothing tangible resulted from the proposal of A.S.L.

In 1931 the American amateur telescope maker Daniel E. McGuire of Shadyside, Ohio, independently proposed the same escape from the fussy difficulties connected with shaping the small hyperboloidal secondary. Early in the same year, unknown to McGuire, H. E. Dall of England had made such a telescope. However, Dall did not reveal the method for calculating the curvature of the primary. Alan R. Kirkham of Tacoma, Wash., revealed the method in this department in June, 1938, although he did not actually make the telescope. Because these two were the first who were known to have done serious work on it, this department then suggested the name Dall-Kirkham for the spherical-secondary compound telescope.

Since that time a modest number of Dall-Kirkhams have been built and have proved satisfactory. No claim was ever made that they are optically superior to the Cassegrainian. They are simply easier to make. The older, more difficult Cassegrainian paraboloid-hyperboloid combination still survives, partly from the momentum of tradition, and perhaps because it has been difficult to collect the fragmentary instructions for making the Dall-Kirkham, scattered as they are in several back numbers of *Scientific American*. Robert Turner Smith of Albany, Calif., prepared instructions that are complete in themselves. He writes:

The compound telescope has an appeal which cannot be denied. The Cassegrainian in particular has advantages both in construction and in observation. Its long equivalent focal length is conveniently folded into a tube only about one fourth as long as that of the Newtonian, an arrangement which results in great stability, less vibration and little overhang of mass beyond the bearings of the mounting. The eyepiece is at the lower end of the tube, more easily accessible and with less sweep than the eyepiece of a Newtonian. In the common focal ratios the Cassegrainian has one-half the length but twice the power of the usual Newtonian. When high power is desired, a lower-powered eyepiece with greater eye relief can be used to obtain the same power where a short-focus eyepiece with uncomfortably close eye relief would be needed with a Newtonian. When good seeing prevails, the maximum useful power of 50 times the aperture in inches can be attained without resorting to an ultra-short-focus eyepiece.

Despite its advantages the true Cassegrainian presents many problems in the polishing and figuring of its mirrors. The primary, although usually an *f*/4 requiring a "strong" paraboloid, is not too much more difficult to figure than an *f*/8 mirror. On the other hand, the convex hyperboloidal secondary is difficult not only to figure but also to test. The high center and turned-up edge is the opposite of what usually "just happens," and the small linear diameter of the secondary for a moderate-sized primary merely adds to the figuring problem. The test of the secondary requires either a flat of the same diameter as the primary or a short-focus sphere of equal diameter for the Hindle test. All this adds up to a project that few amateurs are willing to embark upon. For those whose determination transcends the difficulties, disappointment usually follows when the perfection of figure of the primary is not equaled in the secondary, for a Cassegrainian is never any better than the figure of its secondary. As a result, the Cassegrainian is maligned and has become something to be avoided.

Several years ago Kirkham and Dall investigated the possibility of leaving the secondary spherical and adjusting the correction of the primary to compensate. Since leaving the secondary spherical amounted to overcorrection, it followed that the primary would have to be undercorrected, which is an easier job than full correction. Parabolizing corrects longitudinal spherical aberration. If the secondary is left spherical it introduces a calculable amount of longitudinal spherical aberration into the system. This spherical aberration is negative, and the secondary would be said to be spherically overcorrected. The primary must therefore contain positive spherical aberration in an equal amount and be spherically undercorrected. All that remains is to determine the exact amount of undercorrection necessary.

The formulas for determining the longitudinal spherical aberration and the percentage of undercorrection are not complex, but they do contain sign conventions which must be strictly adhered to, and are therefore subject to error. They can be simplified into the single formula shown at right center in the drawing [*below*]. In this all quantities are considered to

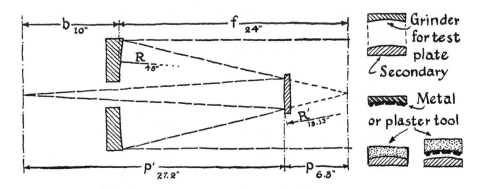

Position of Secondary

$$p = \frac{f+b}{A+1} = \frac{24'' + 10''}{4+1} = 6.8''$$

where A = Amplifying power.

Radius of Curvature of Secondary

$$R' = \frac{2p'p}{p'-p} = \frac{2 \times 27.2'' \times 6.8''}{27.2'' - 6.8''} = 18.13''$$

Percentage of Undercorrection

$$N = 1 - \frac{4p^2}{RR'}\left(\frac{p'+p}{p'}\right)^2$$

$$= 1 - \left(\frac{4 \times 6.8'' \times 6.8''}{48'' \times 18.13''}\right) \times \left(\frac{27.2'' + 6.8''}{27.2''}\right)^2$$

$$= 1 - (.21254 \times 1.5625)$$

$$= .668 \text{ or } 66.8 \text{ percent}$$

Figure 4.1 Rule and example for planning the Dall-Kirkham telescope

be positive and no sign convention errors are likely. N is the fraction of r^2/R for any zone, R is the radius of curvature of the primary, R' is the radius of curvature of the secondary, and p and p' are the conjugate focal distances of the secondary, as shown in the upper part of the drawing.

To clarify the percentage calculation let us take as an example a 6-inch spherical-secondary Cassegrainian in which the primary has a focus of 24 inches, an amplifying ratio of 4, with the focus falling 10 inches behind the surface of the primary. The radius of curvature of the primary is 48 inches and the radius of curvature of the secondary is 18.13 inches, p is 6.8 inches and p' is 27.2 inches. Substituting these figures in the formulas we have the example worked out there.

Now that we are armed with the information on the percentage of correction necessary in the primary, the question arises: How do we produce a good convex sphere? The problem is relatively simple compared with the hyperbolic secondary of the conventional Cassegrainian, both in figuring and testing. The best means of testing a convex sphere is with a concave spherical master, or test plate. When only one convex surface is to be produced the test plate should be one that is quick and easy to make. The glass grinder on which the secondary mirror has been ground need be given only a quick shine, and a suitable test is at hand. Being concave, it can be tested directly by knife-edge and checked for radius with a steel tape. It need only be polished front and back sufficiently to be seen through for the interference tests. In fact, the shorter the polishing period on the concave sphere, the more likely that the curve will remain truly spherical, provided it received a good grind.

Since the grinder has been used for the test plate it will not be available for making the polisher. The best substitute is a concave metal tool, which can be turned to shape in the lathe. A truly spherical and smooth surface is not necessary, since the surface will be covered with pitch in making the polisher. As an alternative a polisher back can be made by greasing the face of the fine-ground secondary mirror, circling it with gummed paper tape and filling this with plaster or dental stone. After the polisher is made and the pitch formed with the mirror, a few minutes of polishing will produce a shine sufficient to obtain the first interference test. This test may show several fringes of difference between the mirror and the test plate. If the fringes are circular and uniformly spaced, then the surfaces are spherical and differ only in radius of curvature, and polishing can be continued. If more than six to ten fringes are apparent, fine grinding was not controlled closely enough to bring the two surfaces coincident and it had best be redone; otherwise prolonged polishing will be necessary to correct the difference.

The final figure of the secondary should be within one-eighth wavelength of truly spherical, but this does not require the same appearance under test as two flats that differ by one-eighth wavelength. In making a flat accurate to one-eighth wavelength there is one and only one surface which is flat, or plano, and until this particular surface is arrived at the flat is not accurate to the tolerance specified. In producing curved surfaces accurate to the same standard there is much greater leeway, since the radius of curvature is not critical and may vary by many wavelengths so long as the spherical surface of the final radius does not deviate from truly spherical by more than one-eighth wavelength. This allows a multiple choice of radii, whereas the flat allows a single choice. Therefore, the spherical surface under test may show as many as four or five fringes and still be classed as accurate to one-eighth wavelength, if no fringe is distorted from symmetrical form more than one fourth of a fringe. (One fringe equals one-half wavelength, one half fringe equals one-fourth wavelength, one fourth fringe equals one-eighth wavelength.)

Testing should be done only after both pieces are thoroughly clean and dusted free of particles which could scratch the surface or hold the pieces apart to cause extra fringes. In the final interference test the test plate should be gently rocked on the mirror so that the center of the fringe pattern moves to all zones of the mirror. In all positions of the circular fringe pattern the fringes should show the same circular form and spacing. Any deviation of one half a fringe is readily apparent if there are fewer than six circular fringes in the diameter of the mirror, but detecting a quarter-wave difference becomes difficult and impossible as the number of fringes increases. The fewer the fringes the more accurately the deviations can be estimated.

In order to observe the fringe pattern easily a moderately monochromatic light source is needed. Fringes appear quite clearly under a fluorescent light, and they stand out even more sharply under a sodium or mercury-vapor light. The light should be diffused, and the lamp is best fitted under the top of a black box with the front open so that the angle of the eye is kept small.

All that remains is the figuring of the convex secondary to a spherical surface. The same techniques are used for correcting convex surfaces as for concaves. Low centers or long over-all radius (concave to the test plate) calls for short strokes or inverted rose laps. High centers or short over-all radius (convex to the test plate) calls for long strokes or straight rose laps. During the polishing-out period appropriate steps can be taken to keep the number of fringes small, and it is quite possible to have a good spherical surface at the same time it is polished out. The secondary need never be tested in conjunction with the primary with which it is going to be used, and large flats or spheres are unnecessary. The primary should be figured and tested to the same degree of accuracy as if it were fully corrected. The allowable error in the accuracy of parabolizing should be adhered to, so that the primary will also meet the one-eighth wavelength criterion.

In calculating the zonal readings for the undercorrected primary it is best to solve r^2/R for the various zones to be tested, subtract to obtain the difference between zones, and *then* apply the percentage figure, obtained from the formula, to each zonal difference amount. For example, let us say that we will test three zones on the six-inch primary for which we determined the percentage correction to be 66.8 per cent, rounded off to 67 per cent. (Three zones will suffice for our example, but more might be desirable for better control of the figure.) We will take zones .75-inch r, 1.75-inch r and 2.75-inch r. For these zones r^2/R will be .012-inch, .064-inch and .158-inch. Subtracting, we get .052-inch for the difference between zones 1 and 2, and .094-inch for the difference between zones 2 and 3. Taking 67 per cent of these figures, we get .035-inch and .063-inch, and these new differences are those to which the mirror is figured. When the zonal test of the mirror shows that it agrees within the greatest allowable deviation from a perfect figure, in this case 5.5 per cent, the mirror will be undercorrected by the proper amount, and it will perform as well with its secondary as a fully corrected primary will with a hyperboloidal secondary.

A complicated ray trace of the spherical secondary would undoubtedly show some higher-order differences of correction when compared with the hyperboloidal secondary system. Coma is probably increased slightly, but for all practical purposes it can be considered negligible. Moreover, most of us are not concerned with higher-order coma or the like. We want a convenient, high-powered telescope that doesn't take 10 years of experience to build. We want to split that double star the book says this diameter should split; we want to find the Great Wall on the Moon, Syrtis Major on Mars, Cassini's division in Saturn's ring, see a transit of a satellite of Jupiter, locate Mercury in the twilight, pick out the Ring Nebula by knowing where to point in Lyra, and accomplish it all with a pair of mirrors that were fun to make in the first place. If, after our apprenticeship on the standard beginner's six-inch Newtonian, we are going to build only one telescope and then see the sights, we might as well make the one that has the highest power and is easiest to use. Even if we just like to make telescopes, the spherical-secondary Cassegrainian is an intriguing one to add to the list. It works fine. I know because I've got one.

5 PRINCIPLES OF ERECTING TELESCOPES

Conducted by Albert G. Ingalls, March 1951

The basic principles of the lens-erecting system for terrestrial telescopes are described in the following notes by Allyn J. Thompson, New York, author of the book *Making Your Own Telescope*.

The manner in which the terrestrial telescope functions is shown in the drawing on the next page. Rays from a distant axial object point are brought to an axial focus in the plane y of the objective O. Rays (dashed lines) from a marginal point of the field are brought to a focus on the opposite side of the axis in the same plane y. This gives us an inverted image of the distant object. Now suppose we regard the erecting lens L as a projecting lens. By placing it at some distance p that is greater than its focal length from the image y, an image of the latter will be projected to y'. The image is turned around or inverted in the course of projection, so that the new image is erect. The distance p' to the projected focal plane y' is found from equation 1 where F' is the focal length of L. The relative size of the images y' and y depends on the ratio p'/p. In the diagram, p' is twice the distance p; the image y has therefore been doubled in the projection, and in effect the focal length of the objective has been amplified two times. The equivalent focal length of the telescope is given in equation 2. The new image y' is viewed with the aid of the eyepiece E, and the total magnification M of the telescope is given by equation 3 where F'' is the focal length of E.

A field stop of suitable aperture should be placed in the plane y' to delimit the field to one of even illumination, and to cut off the poorly imaged external parts. A reticle can be used either in the same plane or at

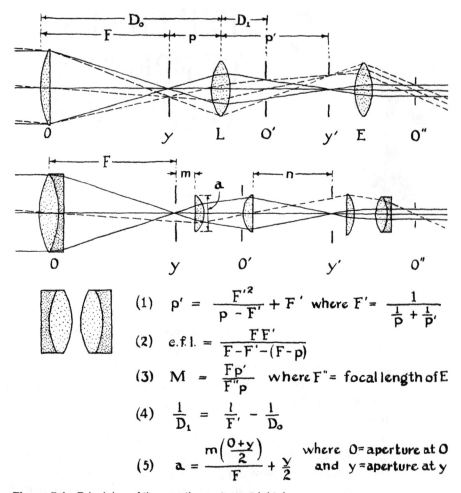

(1) $\quad p' = \dfrac{F'^2}{p - F'} + F'\quad$ where $F' = \dfrac{1}{\frac{1}{p} + \frac{1}{p'}}$

(2) $\quad e.f.l. = \dfrac{F\,F'}{F - F' - (F - p)}$

(3) $\quad M = \dfrac{F p'}{F'' p}\quad$ where $F'' =$ focal length of E

(4) $\quad \dfrac{1}{D_1} = \dfrac{1}{F'} - \dfrac{1}{D_0}$

(5) $\quad a = \dfrac{m\left(\dfrac{O + y}{2}\right)}{F} + \dfrac{y}{2}\quad$ where $O =$ aperture at O and $y =$ aperture at y

Figure 5.1 Principles of the erecting or terrestrial telescope

y. By mounting the erector lens *L* in a separate focusing tube so that the distances *p* and *p'* can be varied, a similar variation in magnification is effected. This action of course will affect the position of *y'*, and complicates the matter of reticle installation. Also, the use of the telescope for different object distances affects the position of both *y* and *y'*. In these circumstances, the best thing to do is to mount the reticle in a unit within which the eyepiece can focus.

Just as the image *y* is projected by the lens *L*, an image of the objective *O* is projected to *O'*. The distance *LO'* is found from formula 4, where D_0 is the sum of the distances *F* and *p*. The relative sizes of *O'* and *O* are proportional to their distances from *L. O'* is the exit pupil of the system *OL*, and is the entrance pupil of *E*. As any light passing beyond the boundaries of *O'* may serve only to fog up the image *y'*, a stop (called an erector stop) should be placed in the plane *O'*. An image of *O'* is projected by the eyepiece to *O''*; this is the exit pupil of the telescope, and it is in this plane that the lens of the eye should be placed when viewing the image. The distance *EO''* is known as the eye relief or eye distance, and from formula 4 it is apparent that this distance is greater than if the erecting lens were absent, a condition that is also found in the astronomical telescope. The diameter of *O''* is equal to that of *O* divided by the total magnification of the telescope, provided of course that there is no curtailment of *O'* by the erector stop.

The erector eyepiece of the upper drawing is the type that was devised by the German mathematician Christoph Scheiner in 1637. In practice it is rendered nearly useless by overpowering aberrations. An immense improvement was effected in 1645 by the Bohemian astronomer Antonius Maria Schyrleus de Rheita, who substituted for *L* two equal plano-convex lenses separated by their focal lengths, with the convex surfaces facing each other, and for *E* a Huygenian eyepiece. This arrangement corrects for chromatic difference of magnification, and at *f*/8 or higher ratios its performance compares well with that of modern designs. Rheita's erector is shown in the second drawing, although a Kellner eyepiece is used in that illustration instead of a Huygens.

In using a two-lens erecting system, the amplification or relative sizes of apertures at *y* and *y'* depend on the proximity of the erectors to *y*. If this distance is chosen so that *m* and *n* are equal, the image sizes are equal, and the magnification is unity. By moving the system closer to *y*, the projection distance and the image size are increased. Actually the projection distance is not represented by *n*, nor the object distance by *m;* these measurements are referred to what is known as the principal planes of the system. There is no need, however, to delve further into the study of optics. The distances *m* and *n*, which are all that are necessary for construction, can be obtained by using an illuminated artificial image at *y*, and experimentally positioning the erectors until a sharply focused image of the desired enlargement is picked up on a ground-glass screen.

With the lens separation given for Rheita's erector, the pupil at *O'* is formed within the second erector lens, so the best place for the erector stop is immediately in front of that lens, as shown. To avoid vignetting in the external parts of the field at *y'*, the erector lenses must be of suitably wide diameter. The clear aperture of the first lens should be as shown in equation 5. The clear aperture of the second lens need be no more than that at *O'*.

The shorter the focal length of the separate erector elements, the less will be the over-all length of the telescope. Modern terrestrial telescopes employ achromatic eyepieces, seldom of more than three inches focal length. Most achromatic eyepieces usually function well for this purpose, as do projecting systems. A single doublet lens will often be found to perform satisfactorily. An arrangement frequently employed is that shown in the third drawing—two identical achromats placed almost in contact. Hardly anything will be found to excel the performance of a Hastings triplet. This is a magnifier designed by Charles S. Hastings. It is known as a triple aplanat. The requisite aperture can be determined from formula 5.

In lieu of an optical bench, experimental setups can be made by standing the various lenses edgewise in lumps of artists' modeling clay mounted on a stick. Care must be taken to maintain a fairly good axial alignment of the elements. In making tests for image quality, magnification, and best lens spacing, the stick should rest securely on some solid object. This allows for painstaking inspection of the image.

A Hastings triplet makes an excellent substitute for a Barlow lens in effecting image enlargement in the astronomical telescope. A lens of 1-inch focal length can be used with reflectors down to about $f/7$. It gives a slightly curved field, which at low amplifications may be conspicuous in some eyepieces taking in a wide linear field. This is easily mitigated, however, either by stepping up the amplification or by using a higher-power eyepiece. With the use of two such lenses placed nearly in contact, field curvature vanishes, and the spherical aberration is so reduced that the pair can safely be used down to about $f/4.5$. (It is assumed in either case, of course, that the mirrors are perfectly corrected.)

6 POWERFUL POCKET TELESCOPES

Conducted by Albert G. Ingalls, April 1952

Many years ago Horace E. Dall of 166 Stockingstone Road, Luton, Bedfordshire, England, built the remarkably compact terrestrial and astronomical telescope shown in Roger Hayward's drawing on the next page. It is a modified Cassegrainian of 3¼-inch diameter, mounted mainly on light aluminum and duralumin parts. It weighs only eight ounces, and can be folded so flat that it juts no farther out of a vest pocket than a fat fountain pen. It gives erect images, and the magnification is continuously variable (pancratic) from 35× to 80× by pulling out the eyepiece. Dall found it a treasure for either day or night use. It has a better light grasp than a three-inch refractor and resolves stars down to Dawes' limit.

Starting from Dall's specifications for diameters, distances, ratios and radii, as shown in the drawing, Frank McCown of Holtville, Calif., has built a similar telescope of four-inch diameter, still small enough to be carried disassembled in a thin 8- by 15-inch box. In the drawing on page 47 McCown's mounting is shown, for convenience, with polar axis vertical. The polar axis is the little stub projecting from a 90-degree angled member in the lower left-hand corner. When this is pointed at the Pole Star in the direction N by adjusting that member, the mounting becomes equatorial.

"This little portable four-inch instrument has performed so well," McCown writes, "as to retire a more cumbersome six-inch Newtonian reflector."

Dall's specifications are shown instead of McCown's so that other builders can work directly from the originals, choosing their own desired sizes. It is sound optics to alter the size of a telescope by reducing or

45

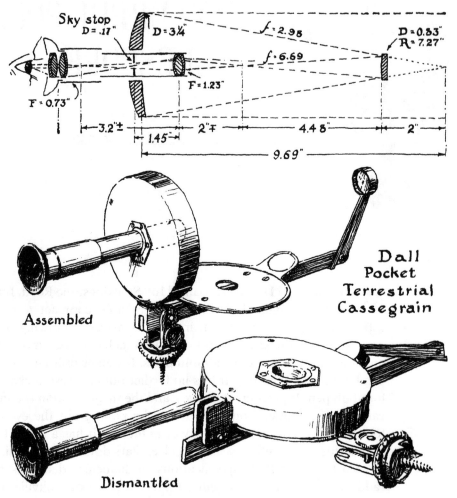

Figure 6.1 A powerful pocket telescope made in Britain

enlarging all measurements by the same proportion. (Preferably not including the indexes of refraction, dispersion and density, or the type number from the glass manufacturer's list.)

The telescope is a modified Cassegrainian of the Dall-Kirkham type with spherical secondary and near-ellipsoidal primary mirrors. The finicky job of figuring the little hyperboloidal secondary of the straight Cassegrainian is eliminated; this advantage is why the Dall-Kirkham is supplanting the old-fashioned type. Design data for the Dall-Kirkham were rounded up in Chapter 4, "The Dall-Kirkham Telescope." Dall figured the primary, an ellipse of eccentricity .873, by the direct focal test.

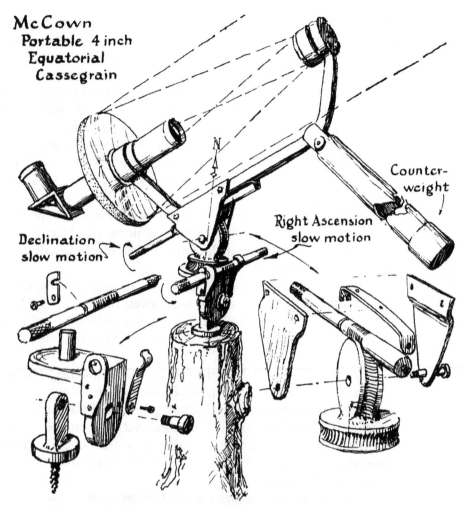

McCown
Portable 4 inch
Equatorial
Cassegrain

Counter-weight

Right Ascension
slow motion

Declination
slow motion

Figure 6.2 In California the pocket telescope grew larger

Just as the Dall-Kirkham is a modification of the Cassegrainian, so Dall's own telescopes are a further modification of the Dall-Kirkham. The modification of the modification which might be called the Dall Cassegrainian, and which Dall regards as a most valuable addition to the spherical secondary, is the placing of an erecting lens between the primary and secondary mirrors.

Some years ago Dall made a 15½-inch telescope of this kind and listed the rewards from adding the little erector. It enables the sky-flooding diaphragm to be moved from the eyepoint (where it is an infernal nuisance, having to be fitted to each eyepiece and impossible to keep adjusted because its aperture is so small) to a position between the erecting lens and the eyepiece, where it is out of the way. It has a large aperture, and therefore is easier to keep aligned. It permits the use of wide-field eyepieces with comfortable eyepoint, greatly appreciated by spectacled observers. An iris diaphragm can be used for the sky stop, permitting the aperture of the telescope to be varied by a lever from full to nothing during observation. The long focus of the Cassegrainian can be shortened. Variability of lens-to-secondary and lens-to-eyepiece distance gives a final image varying in angular aperture (on the 15½-inch) from $f/10.5$ to $f/25$, permitting continuously varying power from 1 to 2½ for each eyepiece. And the final image is erect.

Dall designed and built cemented triplet erectors for his two telescopes. The one for the 3¼-inch gives uniform zonal focus at mean cone focus 1.23-inch, with .62-inch diameter. The focal ratio of the 3¼-inch is variable from 7.5 to 18.

The symmetrical, two-doublet form is more commonly used. Data for designing two-doublet erectors were given in Chapter 5, "Principles of Erecting Telescopes." McCown's erector's combined focal length is 1.9 inch. He found it satisfactory, especially since less than its full aperture is used.

"I am enthusiastic about the Dall erecting telescope," McCown writes, "as the erector serves the purpose of a Barlow lens, also making possible a smaller secondary. The self-collimating feature of the centrally supported mirror has been successful. The threaded ring on the large eyepiece-erector tube screws against a cork ring at the back of the mirror and holds everything in alignment. Once it is collimated, no flare is visible even after repeated takedowns.

"My midget slow-motion mounting is far from perfect. The rods interfere with one another in some positions, but this can be quickly remedied by reversing the base. All but a small part of the sky is accessible. The pressure springs have enough give to allow either rod to be pushed out of gear for a change of view."

In the tubeless telescope, fogging of the mirrors from the observer's breath may be avoided by selecting only breathless views. In McCown's case, which may be unique, this is facilitated by the fact that his farm is below sea level in the Salton Sea depression, so that the instrument is a submarine telescope.

7 A NOVEL REFRACTING TELESCOPE

Conducted by C. L. Stong, May 1958

Familiar optical devices such as eyeglasses and binoculars form an image by refracting light with glass lenses. Yet when an amateur makes a telescope its main optical element is almost always a mirror. Why? Primarily because in making the mirror of a reflecting telescope the amateur need grind and polish only one optical surface. If a refracting telescope is to bring images in various colors to even roughly the same focus, its objective lens must ordinarily consist of two pieces of glass. Thus the maker of a refractor objective must grind and polish four surfaces. To make matters worse, the edges of both lens elements must be ground and fitted to a precisely machined cell so that their curves are centered on the optical axis of the telescope.

J. H. Rush of Boulder, Col., suggests an easier way to make a refractor: a remarkable design which, although it is completely corrected for color, has an objective lens consisting of only one piece of glass!

Rush writes: "Chromatic aberration has plagued the designers of refracting telescopes from the time of Galileo down to the present. This defect causes images to be blurred by overlapping colors and to be surrounded by colored halos. Yet a truly achromatic (color-free) refractor design has been available since 1899. In that year a German optical worker named L. Schupmann published a small book on what he called 'medial' telescopes. His work was curiously neglected. Lately it has been rediscovered and adapted to modern telescope designs by James G. Baker of the Harvard College Observatory, to whom I am indebted for much of my information on the Schupmann system.

49

"A brief review of the evolution of refractors may focus some light on the advantages and disadvantages of the Schupmann telescope. Isaac Newton decided (on the basis of inadequate experiments) that 'all refracting substances diverge the prismatic colors in a constant proportion to their mean refraction'; that is, the focal lengths of a lens for two particular wavelengths of light must always be in the same ratio, regardless of what kind of glass the lens is made of. This conclusion was in error, but it led Newton to decide that a color-compensating lens was impossible because any two pieces of glass that could cancel the dispersion of colors would also cancel the desired refracting power. He turned his attention to reflectors, and his great reputation discouraged others from challenging his conclusions.

"About 1730 Chester M. Hall of England disputed Newton's authority and produced the first achromatic lenses. It is interesting to note that Hall was encouraged in this undertaking by his observation that the various 'humours' in the eye produce an achromatic image. He concluded that a combination of different glasses should also do so. His observation was just as erroneous as Newton's (the eye is not achromatic), but Hall's error was optimistic and led to a notable advance in optics.

"Hall did not do much with his invention, but a few years later John Dollond independently invented the achromat and promoted it so energetically that it revolutionized telescope design. Achromatic lenses take advantage of the differing optical properties of various kinds of glass. A converging (positive) lens, usually of crown glass, is combined with a diverging (negative) lens, usually of flint glass. Because the dispersion of the flint glass (its power to bend light of one color more than another) is greater than that of the crown glass, the flint element can be designed to reverse and cancel the color dispersion of the crown element without entirely canceling its mean refractive power. The resulting 'achromatic doublet' is commonly used as a telescope objective, and the same principle is of course applied in more complex lens systems.

"Unfortunately the doublet does not give perfect color correction. Color dispersion by glass or other substances is irrational. That is, the refraction for different colors is not proportional to their wavelength. Hence when refraction is plotted against wavelength, the resulting curves for different glasses do not have the same shape. A doublet can be designed to bring two chosen wavelengths to a focus at exactly the same distance from the lens; intermediate wavelengths fall a bit short, and those beyond the corrected region deviate seriously from the common focus. This 'secondary spectrum' is quite troublesome in a large instrument. The 40-inch achromat at the Yerkes Observatory, for example, focuses the yellow-green about a centimeter nearer the lens than the red and blue for which it is corrected. If three different kinds of glass are combined, three wavelengths can be brought to a common focus. Such a lens is called an apochromat. It gives a much better color correction than does the achromat, but is costly and suffers from other disadvantages.

"In complex lens systems an additional defect called lateral color usually appears. Such a system may focus all wavelengths of interest at practically the same distance from the last lens element. Yet the images in different colors may be of different sizes, so that they do not coincide except on the optical axis of the telescope. The result is an overlapping color effect similar to that produced by the more familiar longitudinal color just discussed.

"At this point one may ask: Why bother with refractors at all? Mirrors are entirely free of chromatic aberration, since the simple law of reflection holds for all wavelengths. The answer is mainly that a mirror system usually wastes more light than a good achromatic objective (a minor objection), and that, usually having only one figured surface, the mirror does not permit adequate control of other aberrations.

"The ideal mirror shape for focusing parallel rays, with which we are concerned in an astronomical objective, is the paraboloid. Its spherical aberration is zero. All light coming into such a mirror parallel to its axis is reflected (within the limitations due to the wave properties of light) to a single focal point on the axis. So far, so good. But parallel rays coming in at an angle to the axis, such as the light from a star that is not centered in the field, do *not* converge to a common focus—as any telescope-making enthusiast is painfully aware. The image is distorted by two types of aberration: astigmatism and coma. In the case of astigmatism, an object point off the optical axis is imaged as two lines at different focal distances and perpendicular to each other. The effects of coma, on the other hand, resemble those of spherical aberration. But instead of a point-source coming to focus as a circular patch of light in the plane of the image, as in spherical aberration, coma results in a comet-shaped patch. Moreover, the image field is a curved surface.

"There is nothing you can do about these defects. You have already determined the shape of the mirror, making it a paraboloid to eliminate spherical aberration; and you have determined its scale, to obtain the desired focal length. You can do nothing more with a single surface. Of course, in the case of compound reflecting telescopes, the experts do some tricks with secondary mirror surfaces and sometimes with special correcting lenses, but most amateurs leave these strictly alone.

"In a simple lens the index of refraction of the glass and the *difference* in the curvature of the two surfaces determine the focal length. By 'bending' the lens—keeping this difference constant while changing both radii of curvature—the designer can eliminate all coma and nearly all spherical aberration. Of course the designer still has to live with astigmatism, chromatic aberration and curvature of the field. In designing a doublet, one computes powers for the two elements that will give the desired focal length and color correction. Then, by suitably bending both elements independently, one eliminates both coma and spherical aberration. The perfection of these corrections depends very much on some subtle wisdom in the choice of glasses, which the designer usually attains by surviving his previous efforts.

"Here, then, is the principal advantage of the refractor. A doublet or triplet lens affords the designer degrees of freedom enough to correct the most troublesome aberrations independently. However, astigmatism remains a problem. It is an obstacle to good definition over wide fields, and usually limits an astronomical refractor to a very small area of the sky. But a reflector is even more sharply limited by astigmatism and coma.

"About 1930 Bernhard Schmidt of Germany came up with an excellent solution to these limitations with his now-famous lens-and-mirror combination. His instrument is especially advantageous for fast photography of wide angular fields at low magnification. Its peculiarly curved lens, or 'correcting plate,' is an obstacle to any but the most skilled opticians, but modified designs using spherical lens-surfaces have eliminated even that difficulty for some purposes. Yet the Schmidt is not a substitute for the long-focus refractor, so useful for astronomical observations that require a relatively large image. A Schmidt telescope is twice as long as a simple reflector or refractor of the same focal length.

"Schupmann described two types of unconventional telescopes: the brachy medial (or brachyt) and the medial. The brachyt is not capable of correcting chromatic aberration completely. For that reason I have not investigated its possibilities in detail. It has the great advantage of compactness and might be capable of acceptable performance over a limited field and spectral range, if carefully designed. The optical path of the brachyt is depicted in the accompanying drawing [*page 53*].

"Schupmann's other design, the medial, is something else again. It is capable of practically perfect correction of chromatic aberration over the full photographic spectral range. The medial, like the Schmidt system, uses both lenses and a mirror, but it is mainly a lens instrument and even in its simplest form it need not be much longer than the equivalent simple refractor. The objective is a simple lens of a good telescopic-quality crown glass, such as borosilicate crown No. 2 (BSC-2) or Schott boronkron No. 7

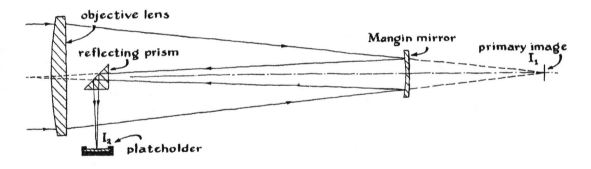

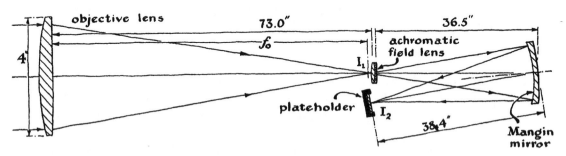

Figure 7.1 The optical path in the two kinds of Schupmann telescope: the brachyt (*top*) and the medial (*bottom*)

(BK-7). It is designed to have the desired focal length for some intermediate wavelength and is bent for zero coma.

"The mirror of the Schupmann telescope is called a Mangin mirror. It is simply a negative lens whose convex back surface is aluminized [*see drawing on page 54*]. It is made of the same glass—preferably the same melt—as the objective.

"The secret of the medial's performance is found in the field lens. Its function is to image the objective onto the Mangin mirror. Optically the effect is to superimpose the objective and the Mangin so that their combined dispersive power is that of a flat plate. The Mangin is bent so that its surface contributes enough positive power to form the final image.

"You can easily see how the medial tends to correct chromatic aberration. Since the lens power of the Mangin is negative, the focal length of the Mangin is greater for blue than for red light. But the primary image formed by the objective in blue light is nearer the objective and thus farther from the Mangin than the red image. Consequently the final image in both wavelengths will tend to fall at or near the same distance from the Mangin, because the difference in object distance is offset by the differ-

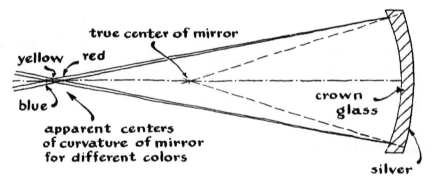

Figure 7.2 A Mangin mirror of crown glass performs like a negative lens of flint glass

ence in focal lengths of the Mangin for blue and for red light. Detailed calculation shows, however, that this correction cannot be exact for more than two wavelengths unless a field lens is used to image the objective onto the Mangin.

"If the field lens were perfectly achromatic, the Mangin could be designed to form final images of the same size and at the same distance from the Mangin in all wavelengths. Practically, the field lens cannot be perfectly achromatic, and its secondary spectrum produces a slight amount of lateral color in the final image. Even a simple, single-element field lens will reduce the secondary spectrum in the final image to a small fraction of that produced by an achromatic doublet objective of equivalent power. A good crown-flint field lens, designed for the image and object distances at which it is to be used, contributes an entirely negligible residue of chromatic aberration over the spectral range and image fields ordinarily used.

"To sum up: The medial telescope uses a negative lens to cancel the dispersion of a positive objective of the same glass, and then interposes a concave mirror to intercept the light that has been so treated and focus it as the final image. Why bother? Why not forget the lenses and just use a mirror in the first place? The complication is justified by the superior optical corrections that can be made in the medial telescope. The medial telescope has several advantages, in addition to freedom from color, over the ordinary refractor. Its spectral range is not limited by the absorption in flint glass because other materials can be substituted—even in the field lens. The surfaces of the objective are less sharply curved than those of the equivalent achromat, so that residual aberrations are reduced and the mechanical strength of the lens is improved. The positive and negative powers of the objective and Mangin result in a nearly flat image-field.

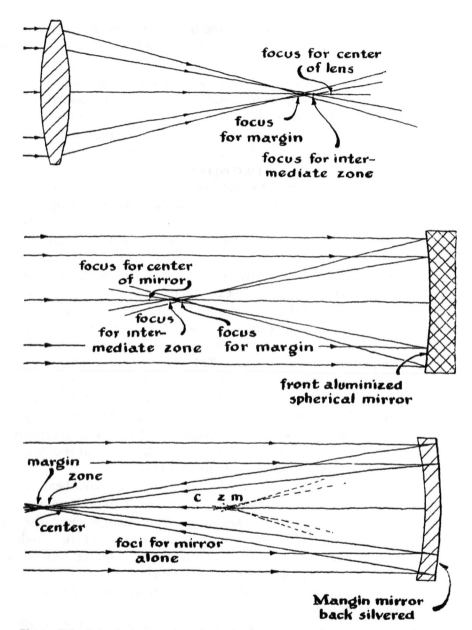

Figure 7.3 Spherical aberration of a Mangin mirror (*bottom*) is the opposite of a simple lens and mirror

"A unique advantage of the Schupmann medial is its adaptability to the Lyot solar coronagraph. The essential features of this instrument are: a simple objective to minimize scattering of white light into the system, a metal disk to eclipse the bright solar disk at the primary image, and a field lens and diaphragms to eliminate white light introduced by diffraction and reflection at the objective. To meet these conditions most coronagraphs have suffered the disadvantages of chromatic images; the medial offers an ideal way to get a color-free image without interfering with the essential optics of the coronagraph.

"Coma causes no trouble, because it is eliminated in the objective by a suitable choice of radii, and in the Mangin by making the image and object distances equal for a mean wavelength. Two difficulties arise, however. Since the shape of the Mangin is totally determined by the twin requirements of color correction and mirror power, it is not possible with spherical surfaces to meet these conditions and at the same time shape the Mangin to cancel the spherical aberration of the objective. Probably the best way out of this difficulty is to figure an aspheric surface on the front of the Mangin, departing just enough from a sphere to bring the spherical aberration of the entire system to zero as determined by knife-edge or other test at the final image. This chore is comparable to parabolizing a mirror objective, but is tricky because of the relatively small size of the Mangin. It is possible, of course, to make the corrections on the primary, which is somewhat easier to work because of its size. Or, if you insist on spherical surfaces, you might make the Mangin lens and mirror elements separately, and bend the lens to cancel the spherical aberration (though I have not checked this possibility in detail).

"The Mangin must be tilted so that the reflected beam will clear the field lens and give access to the final image. This tilt introduces additional astigmatism into the image. Hence the tilt angle should be held to the absolute minimum. In a large instrument, astigmatism introduced by the tilt is eliminated by figuring a toroidal surface on the back of the Mangin, but it is more practical in a small telescope to cancel the astigmatism by the stratagem of tilting the objective on an axis perpendicular to the tilt axis of the Mangin.

"To indicate what can be done, I have computed the objective and Mangin for a system of aperture $f/18$ at the wavelength of the hydrogen alpha line of the spectrum (6,563 angstrom units). This system is suitable for amateur use. The plans are based on an objective aperture of four inches. They can, of course, be scaled to a larger size, if the proportions among all dimensions are maintained. Data for the objective and the Mangin are given in the accompanying table [*next page*].

	OBJECTIVE	MANGIN
MATERIAL	BSC-2	BSC-2
RADIUS (FRONT SURFACE)	40.44 INCHES	–9.2 INCHES
RADIUS (BACK SURFACE)	–440.8 INCHES	18.635 INCHES
THICKNESS (CENTER)	.5 INCHES	.25 INCHES
CLEAR APERTURE	4.0 INCHES	2.0 INCHES

Table 7.1   Data for the objective lens and Mangin mirror of a Schupmann telescope

"In the design presented here the final image will be about 38.4 inches from the back of the Mangin. A diagonal eyepiece will be necessary to keep the observer's head out of the optical path. The system will be about 9.5 feet long—awkward, but not impractical, since the elements are light and little more than self-support is required of the tube. The telescope can of course be shortened by folding the optical path with a double mirror or reflecting prism behind the field lens. In addition to compactness, this arrangement has the advantage of locating the eyepiece in the usual position at the rear of the telescope.

"The Schupmann medial principle offers an opportunity to make a telescope of exceptionally fine definition, and making the telescope is not particularly difficult for the advanced amateur. Only one unconventional element is used, and it requires only one aspheric surface—a job no worse than parabolizing an ordinary mirror."

8 "OFF-AXIS" REFLECTING TELESCOPES

Conducted by Albert G. Ingalls, February 1954

Obstructions in telescope tubes, such as diagonal and secondary mirrors and their mechanical supports, not only cut off light but also cause injurious diffraction. Diffraction vitiates the image and reduces contrast of fine detail, such as lunar and planetary features. The radial spikes that project from images of bright stars on some photographs are evidence of diffraction caused by telescopic obstruction. In general, reflecting telescopes, most of which have central obstructions, give images with lower contrast and slightly poorer definition than those of high-quality refractors.

The late planetary astronomer William H. Pickering described a diffraction experiment which emphasized the impairment of seeing by obstructions. First he deliberately increased the diffraction effect by enlarging the obstruction area of the diagonal mirror with a paper mask extending beyond the mirror. Then he entirely eliminated the obstruction of the diagonal mirror and its supports by another device—a mask which let the incoming rays of light reach the mirror only through a relatively small off-center opening. In spite of the reduction in the amount of light that reached the mirror, the seeing was greatly improved. Adopting this working principle, A. E. Douglass, another planetary astronomer, masked the aperture of a 36-inch telescope in such a way as to leave uncovered only an off-axis circle 13 inches in diameter [*see drawing at upper left in illustration on page 59*]. The result was a net gain in visibility of fine detail despite the loss of brightness.

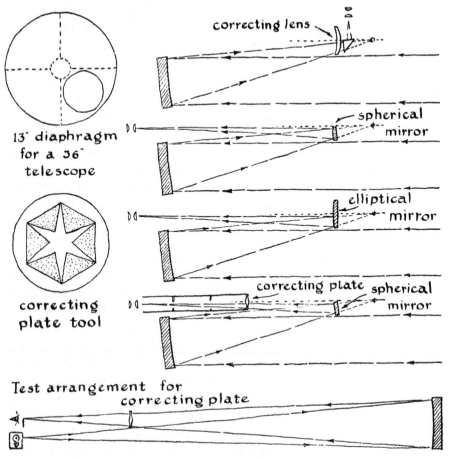

Figure 8.1 Concerning reflecting telescopes with unobstructed tubes

Astronomers have noted these evil effects of central obstruction for many years, but it is not easy to build large off-axis telescopes to avoid them. In recent years advanced amateur observers of the planets and moon who have exhausted the fullest powers of conventional obstructed telescopes have been inventing new telescope designs for reducing diffraction effects. One of these is J. S. Hindle of England, son of the late amateur astronomer J. H. Hindle. He writes:

"When conventional types of reflecting telescopes are used visually on planets, especially with large apertures, the finest detail is usually best seen by using an off-center diaphragm to cover the secondary mirror and its supports, as described by Pickering. However, the arrangement is far from perfect, as the large block of glass in the primary mirror is rarely in a state of thermal equilibrium and definition consequently suffers. An alternative for obtaining an off-axis section of a paraboloid is to coat a spherical mirror with an uneven layer of aluminum. Excellent in theory, this is open to the objection that the thickness of the aluminum deposit must be extremely accurate to get optimum results. The Herschelian reflector, used with a meniscus correcting lens in the return beam [*see drawing at upper right in illustration on page 59*], seems to offer an answer. Moving the lens to or from the mirror gives under- or over-correction for spherical aberration. Although the scheme gives remarkably good results with moderate apertures, it proves cumbersome with large instruments, the observer perforce being perched high on a ladder or platform in the dark. The same dangerous and uncomfortable situation arises when an off-axis paraboloid is used.

"The skew Cassegrain was first used to avoid central obstruction by Karl Fritsch of Austria, who made many small ones about six inches in diameter, and A. A. Common of England, who made a 12-inch from which he claimed to get good visual results. These instruments used two spherical mirrors [*see cross section second from top in illustration on page 59*]. Although the aberrations of the secondary tend to cancel the spherical aberration of the primary, the method fails when tried on large apertures, the residual spherical aberration being such that critical detail is unobtainable under high magnifications. Tilting the secondary tends to minimize spherical aberration but introduces astigmatism, which causes star images to appear as short straight lines.

"Another modification which at one time looked promising was an elliptical secondary about two thirds the diameter of the primary [*see cross section third from top*]. This large secondary was necessary because only an off-center section of it was used. The scheme failed on large sizes, due to flexure of the secondary, which could be supported only at the edge. The elliptical secondary also is very difficult to test and figure.

"To dispose of these defects of the off-axis telescope I have designed and constructed a Cassegrain-Schmidt combination [*see cross section fourth from top*]. It has two spherical mirrors and a small correcting plate similar to a Schmidt plate to cure spherical aberration. In sizes larger than about eight inches this type will give far sharper detail and stand far higher magnification than the conventional Newtonian and Cassegrain types. My 12-inch will stand 600 diameters on most nights and still give an image nearly as sharp and crisp as when using only 60 diameters. Tested against three different 12-inch reflectors and one 15-inch, it has easily outclassed the performance of them all when planetary detail was being observed. The data for a 12-inch telescope of this type are as follows:

Focus of primary, 10 feet or more. Diameter of convex, 3 inches. Radius of curvature for convex, 62 inches. Cone cut off by convex, about ⅓ focus. Diameter of correcting plate, 3 inches; thickness, about ¼ inch.

"If these proportions are approximated, the primary mirror may be used for testing the figure of the correcting plate. The angle of the return cone of rays, when an illuminated pinhole is placed at the center of curvature of the primary, will be identical with the angle of the cone from the secondary when the finished instrument is used at infinity. The amount of deviation from flatness that must be imparted to the correcting plate can be calculated from the familiar formula r^2/R, with one important difference. The r in this case does not refer to the semi-diameter of the primary; it refers to the distance from the optical axis of the system to the farthest point on the circumference of the primary.

"When testing the correcting plate the spherical primary is set up as usual for the Foucault test but using about six inches' separation between the pinhole and knife-edge [*bottom drawing*]. The plate is then placed between the primary and knife-edge, at such a distance that it just fills with light when viewed from the center of curvature, without any of the mirror being visible outside the plate. In the example given the plate must be figured until the primary assumes the appearance of a hyperboloid, with the outer rays coming to a focus about ⅞ inch longer than the central rays. It will be necessary to attack both sides of the plate to obtain this deviation, if the plate is made from polished plate glass. It is advisable to cut several square pieces and test for astigmatism in the return cone of rays from the primary at center of curvature. The best shape of polisher for the plate is shown in the illustration. If the plate is mounted on a rotating spindle, it may be quickly roughed to shape with the ball of the thumb, and afterward finished off with the polishing pad.

"If the plate shows traces of astigmatism when finished, do not scrap it but rotate it in different positions when in the telescope, as only half of the plate will be in use and a good diameter usually can be found.

"The best test for the convex is King's, but a fair idea of the figure may be obtained by placing the pinhole at four times the focal length of the primary away from the primary and intercepting the return cone of rays with the convex, so as to reflect the cone back to the edge of the primary. On passing a knife-edge across the apex of the cone, an even darkening of the light should be observed. The big snag to this method, of course, is that a long testing room is required. The primary should be aluminized for the test, because there are two reflections.

"A slightly positive meniscus lens may be substituted for the correcting plate, but it will give a rather curved field of view. The chromatic aberration caused by the correcting plate is so small as to escape detection with quite high powers.

"The stops shown between the eyepiece and the plate are very important, as objectionable sky flooding will occur if they are missing or imperfectly positioned.

"Moving the correcting plate nearer the convex increases correction for spherical aberration, and moving it away has the reverse effect, hence good definition will not be attained until this adjustment has been made perfect.

"To collimate the instrument it is necessary to place a button or small circular object exactly in the center of the primary, at the open end of the tube. Both mirrors are then adjusted until the reflection of the mirror and the button are seen concentric in the convex, looking from the eyepiece drawtube with the eyepiece removed."

The off-axis Cassegrain shown in the second drawing from the top was invented in 1876 by I. Forster and Karl Fritsch of Vienna and was known as the brachyte. In the December 1, 1952, issue of *The Strolling Astronomer*, organ of the Association of Lunar and Planetary Observers (ALPO, *www .lpl.arizona.edu/alpo/*), Guenter Roth and E. L. Pfannenschmidt described a modern German variation called the neobrachyte. Roth's *f*/20 "neo-bra" has an eight-inch spherically concave *f*/12 primary mirror, its axis tilted to an angle of 3 degrees, 16 minutes and 30 seconds of arc with the incoming light. The four-inch *f*/24 spherically convex secondary mirror has the same radius of curvature as the primary and is tilted to an angle of 12 degrees, 56 minutes from the new or deflected axis of the primary (though the three axes involved are not in the same plane). The secondary is separated from the primary by a distance of seven times the diameter of the primary. Astigmatism is eliminated by deforming the secondary to a slight cylindricality with a screw pressing against its back along one diameter. This subterfuge, which avoids the necessity of figuring an aspherical correcting plate, is the secret of the neo-bra.

According to the authors an eight-inch neo-bra will show perfectly round, concentric star images at powers of 300 to 450. Larger sizes require a shorter focal ratio, ray tracing and a Dall-Kirkham primary. (For more information on Dall-Kirkham telescopes, see Chapter 4.) Excellent star images were given by a 12-inch neo-bra at a power of 500.

Roger Hayward, who made the illustration from Hindle's sketches, comments: "I recall J. A. Anderson (who had charge of the optical parts of the 200-inch Hale telescope) saying that the ideal size of telescope for visual observation was around 10 inches. In a smaller telescope the image suffers from too little resolving power, in a larger one the seeing deteriorates. This is because atmospheric waves that mess up the air commonly are of such dimensions as to make the image from a 10-inch disk wobble about at a rate which the eye can follow. At twice this size or larger the image will go in and out of focus as the light from two sides of the disk is refracted together or apart. I feel that this, more than the absence of obstruction, is likely to have been one reason why the 13-inch diaphragm opening improved the seeing with the 36-inch telescope."

Much has been written about the evil effects of central obstructions on visual observation, but there is no agreement about the effects. Laboratory experiments which exclude the effects of the turbulent atmosphere do not agree closely with tests at the eyepiece out-of-doors. The laboratory tests seem to prove that there is no serious impairment of image quality until 20 to 30 per cent of the diameter of the aperture is obstructed. On the other hand, E. K. White of British Columbia found at the eyepiece that even the 17 to 25 per cent obstruction in the typical Newtonian telescope caused pronounced diffraction effects on the images of planets. With a reduced diagonal which obstructed but 10 per cent of the aperture diameter, he found that the image quality was greatly improved. The editor of *The Strolling Astronomer,* in which the experiment was described, remarked that White had observed little-known delicate features on Saturn about as well with a nine-inch telescope as he had been able to do with an 18-inch telescope.

Anderson has described a simple method for observing atmospheric waves that impair good seeing. These waves are believed to be ripples at the interfaces between moving layers of warm and cold air high aloft. Anderson points a small low-powered telescope—for example, with one-inch aperture, eight-inch focal length and a low-powered eyepiece—at a bright star, holding the eye end with one hand, the objective end with the other. Then he swings the objective end in a circle about a quarter inch in diameter at the rate of four or five revolutions a second. Due to the persistence of vision, the star image is drawn into a nearly complete circle if the rate of rotation is correct. You soon learn to judge the state of the seeing by the number of bright patches in the circle. The bright patches correspond to the moments when the concave crest of the wave is in line with the telescope so that it refracts the sides of the interrupted disk together, and the darker portions correspond to the convex trough of the waves, which Anderson assumes from experience are about six inches in length. In good seeing the frequency of the bright patches with this scintillometer ranges from a few to 25 or 30 a second, while in bad seeing there are often 150 a second.

Ed.

Does the factor described by Hayward—the length of the atmospheric waves high aloft—have anything to do with this equality? David W. Rosebrugh, a veteran observer, believes that it does. He thinks that the maximum useful aperture for visual use is six inches, because larger telescopes are more sensitive to the atmospheric waves.

After reading a score of articles on the subject, as I have just done, you may come to feel that only two facts have been isolated: (1) unobstructed telescopes are better than obstructed ones; (2) the effects of obstruction are confused by the effects of size of aperture.

9 STEADY TELESCOPE MOUNTINGS

Conducted by Albert G. Ingalls, July 1955 and December 1954

To be functional, any telescope requires a rock-solid mount. This is particularly true for celestial photography, variable star observing, the precise timing of lunar occultations and so on. Knife-sharp photographic negatives and clear photoelectric recordings cannot be made with an instrument that jiggles.

In general the problem of achieving the desired stability has been attacked in two ways. The first reduces the difficulty by exchanging a pronounced slow wobble for a less-pronounced fast one. Low amplitude is bought at the price of high frequency. Axes are made heavier and overhang is reduced to a minimum. In other words the "pendulum" is stiffened and shortened. It is possible to increase the rate of vibration of six-inch instruments to as much as 10 cycles per second by using stubby shafts for axes and appropriately light materials in the construction of the tube and its accessories. However, even this relatively high frequency is well within the range of a good photoelectric recorder. It can easily mask the subtle features of occultation recordings. Thus most designers tackle the problem by the second method: increasing the effective diameter of the axes. A now-classic example of this type of design is represented by the Springfield mounting introduced in 1920 by the late Russell W. Porter. But even though the Springfield axes are defined by plates as broad as the telescope's mirror, physicists of the Army Map Service found it necessary to substitute oversized ribbed castings before the design would yield useful occultation records. Perhaps the simplest way of increasing the effective diameter of axes is to adopt one of the numerous yoke designs. They are cumbersome, of course, and troublesome to move. But if you are willing to sacrifice portability for stability of operation, a good yoke mounting is your dish.

W. P. Overbeck, of Aiken, S. C., has built one that he swears can function simultaneously as a trapeze and a precision photometer. He built the mounting for far less than the patterns alone of a Springfield would cost.

"My enjoyment of astronomy," he writes, "lies in calculating positions of objects or timing astronomical events and subjecting the results to accurate measurement. This provides relaxation from other activities which are not capable of such precise evaluation. Of course the professional astronomer does not define astronomy in such simple terms; he is most interested in phenomena which are very difficult to measure. But the amateur can take delight in finding things within a small fraction of a degree of where he expects to find them or in timing events within a few seconds, particularly when he has built his own instrument.

"To satisfy the desire for accuracy, I soon came to the conclusion that it would be necessary to build a telescope which could be permanently, and very solidly, mounted outdoors. It was difficult to see how reproducible measurements could be obtained with a portable telescope. Secondly, having meager facilities for precision machining, it became clear that I must use a type of mounting in which the principal axes are supported on widely spaced bearings so that any play in the bearings would have a minimum effect on angular motion. Finally, a permanent outdoor mounting must be weatherproof unless one wishes to add the cost of building an observatory complete with dome."

The answer to all these requirements is shown on page 67. Writes Overbeck: "The telescope tube and its declination axis are mounted in a yoke with bearing points spaced about a foot apart. The yoke is similarly mounted between two heavy A-frames with bearings about eight feet apart. With a 'slop' of less than .0005 inch in the bearings, the declination axis is fixed to within .0024 degrees and the polar axis to within .0003 degrees. The A-frames are fastened down with heavy bolts set in a four-inch concrete slab. The arrangement is so solid that one can climb on the frame while making observations. The principal timbers of the mounting are four by six inches. They were measured and cut with considerable care to obtain the correct angle of tilt for the latitude of the site. Each piece was then thoroughly shellacked and painted and finally assembled with half-inch lag screws, most of which were six to eight inches long. Preliminary alignment of the slab and frames was done by first establishing a true north-south line. This was accomplished by suspending a long plumb line

Figure 9.1   A steady telescope mounting made by an amateur

from a temporary support and marking off the position of its shadow as the sun passed through the meridian. By referring other measurements to this line, it was possible to get everything lined up to within a small fraction of an inch and minimize the more precise alignment work required later.

"The yoke was built of half-inch plywood, heavily reinforced with two-by-six- and two-by-four-inch pieces at the points requiring strength. The sides of the yoke, as well as the main A-frames, are built so that they provide 'bearing boxes' at the points where the bearings are to be located."

The detail at the lower left on page 67 shows the arrangement of bearing, clutch and driving gear at one side of the telescope tube. The bearing is made up of three-quarter-inch pipe fittings which were machined down to get a good fit and alignment. Starting from the right, the detail shows a partial section of the tube, which at this point has a double wall for strength. A flange is fastened to the tube and holds a tubular bearing shaft made from a six-inch by three-quarter-inch pipe nipple. The shaft passes through a second flange which is fastened to a two-by-four-by-eight-inch wood block which fits snugly inside the bearing box and is held in position by two lag screws. The lag screws pass through clearance holes in the sides of the box, permitting a small amount of adjustment. On the outer end of the bearing shaft is a third flange, faced with rubber, which forms half of a disk-type clutch. The other half of the clutch is separately assembled with the driving gear on a steel bushing which slides into the outer end of the tubular bearing shaft.

Overbeck continues: "To control the clutch, a long quarter-inch steel rod, having a collar fastened to it, extends through the center of the entire assembly and is threaded into a steel plug at the inner end of the bearing shaft. In the actual assembly the two clutch faces remain in contact. They are surfaced with thin, smooth gasket rubber and require very little release of pressure to change from a locked condition to one which permits free motion. The drive for both declination and right ascension is provided by miniature electric motors which are geared 500,000 to 1 and are housed in the bearing box along with the clutch and gear.

"This makes it easy to maintain good alignment between driving gear and shaft. The box is sealed by a rubber strip which is tacked inside a four-inch hole in the face toward the telescope and which fits snugly against the side of the telescope. A second seal closes the opening around the clutch control rod. It took much longer to invent this assembly than it did to build it. One of its greatest advantages is that it can easily be taken apart, piece by piece, from the outer face of the box.

"The assembly above described is duplicated at one end of the polar axis. The other two bearing assemblies have no clutch or gear but simply have a graduated circle fastened to the flange which forms the inner clutch-face. The outer covers for these bearing assemblies have glass inserts."

The telescope tube is made of half-inch plywood reinforced by several internal "ribs" which are pieces of plywood seven inches square with six-inch holes cut in them. When the tube is not in use, both ends are covered with simple plywood lids such as the one shown in the detail at upper right in the drawing.

A plywood box for carrying the eyepieces is also shown in the drawing. It was carefully built to the same dimensions as the tube. The mirror assembly can be used as a lid for the box when storing the optical parts indoors, or may be quickly installed in the end of the telescope when one wishes to make observations. The diagonal mirror and spider assembly at the other end of the tube is well protected from the weather and remains installed at all times.

Says Overbeck: "After final assembly, precise alignment of the telescope becomes a simple matter. First, the optical axis of the mirror was aligned with the dimensional axis of the tube. This was done by careful sandpapering of the end of the tube that supports the mirror. The alignment is checked by looking through the tube at the mirror from a distance about equal to twice its focal length. From this point one can see an enlarged image of the pupil of one's own eye which, when the mirror is aligned, will be neatly centered on the mounting spider of the diagonal mirror. (My mirror has a small hole in it, when the diagonal is removed, which makes this test very precise.) Second, the polar axis may be aligned by following a star, preferably one near the celestial equator, across the sky and observing apparent changes in declination. This was repeated several times until no apparent change could be seen. Finally the declination axis was checked by making several observations of stars of varying declination to find if there were consistent errors in relative values of right ascension.

"The first adjustments were made by calculating the required motion of various supports and bearings and later adjustments were made 'by feel' during observation. The procedure was successful in bringing the alignment of both axes within about .03 degree of perfect positioning as determined by averaging many individual measurements. This is much better than the precision with which it is possible to read the declination and hour circles, so for practical purposes it is more than adequate.

"Thus the final result is an inexpensive instrument which is capable of precise measurements, which does not suffer from vapor condensation and thermal convection currents as most metal telescopes do. It should be

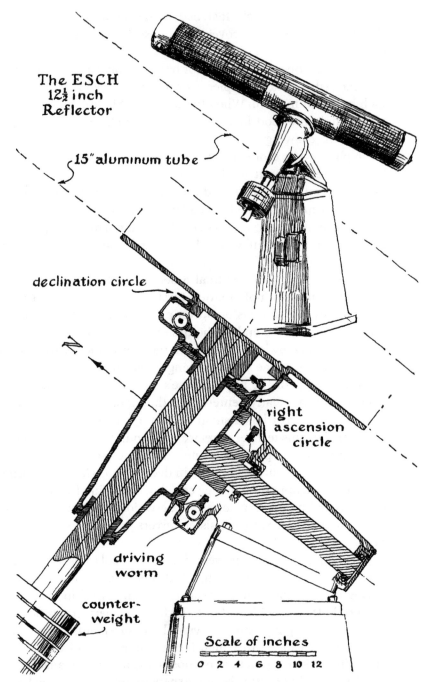

The ESCH
12½ inch
Reflector

15" aluminum tube

declination circle

N

right
ascension
circle

driving
worm

counter-
weight

Scale of inches

0 2 4 6 8 10 12

Figure 9.2 An amateur's telescope re-equipped with adequate axes.

realized, however, that it cost a great deal of time and effort because its precision is largely due to its heavy structure and to the careful attention given to measuring, cutting and finishing of each piece of wood to insure both accuracy and long-term stability of dimensions. For me the effort was well worthwhile in meeting the particular objective I had in mind."

Another stable mount was devised by Robert and Karl Esch of Cherry-vale, Kan., for their 12½ inch telescope. When they put the big telescope to use, the stars danced in the eyepiece. They had used solid steel axis shafts 1⅞ inches in diameter, but evidently these were not massive enough. After inspecting a professionally built instrument at the University of Kansas and noting its massive solidity, Robert went home and wholly redesigned and rebuilt the mounting. Roger Hayward's drawing on page 70 shows its new proportions. Karl says: "This one really has rigidity. I had little to do with the mounting, which was designed and built by my brother Bob, though I made a new 12½-inch mirror for it and was responsible for the rest of the optical parts."

What the brothers had overlooked in their first mounting was the fact that a telescope magnifies tiny vibrations in proportion to its own magnification and must therefore be much more rugged than other machines.

The Esches found that their neat one-legged support for the diagonal mirror resonated in vibration when the Kansas "zephyrs" whistled down the telescope tube, blurring the image. They substituted a more conventional three-legged support spider.

Robert writes: "We are most proud of the simplicity of the driving controls, now that we have added a drive. The drive uses a synchronous ball-bearing motor of ⅒ horsepower and 1,800 revolutions per minute, geared through a 96-3-3-30-100 gear combination to give a ratio of 2,592,000 to 1. While this does not give perfect sidereal time, it would be wasteful to buy special gears for a closer approximation." Karl adds: "I set the telescope on Sirius and went indoors for an hour to blot up heat and on returning I found Sirius still in the field."

10

TWO AMATEUR-BUILT RADIO TELESCOPES

Conducted by C. L. Stong, February 1962

Amateurs traditionally make radios and telescopes, yet it appears that few of them make radio telescopes. Although it may seem a daunting challenge, radio telescopes can be made by any experienced ham with a passion to explore the cosmos. This article describes two amateur-built radio telescopes. One was built by Lyndall McFarland of Winston-Salem, N.C., and the other at Manhattan, Kan., by Walter Houston, Clifford Simpson and Ben Mullinix. The two instruments are comparable in performance but differ in design: one is a reflector and the other a diffractor. McFarland's instrument picks up signals from any given direction by means of a 15-foot paraboloid of aluminum and focuses them on a simple dipole antenna. The Kansas instrument uses a 12-element Yagi array, a series of dipoles supported by a spar that resembles an overgrown television antenna. The length and spacing of the dipole elements were chosen so that radio waves arriving from all but the desired direction interfere, whereas those from the desired direction add constructively at the location of one dipole that feeds a radio receiver.

Both antennas are steerable in altitude and azimuth and have detected the sun as well as the more energetic radio sources in Sagittarius, Cygnus, Cassiopeia and Orion. The resolving power of McFarland's telescope is about 11 degrees of arc; it detects the sun as being a disk some 20 times wider than it appears to the eye. The resolving power of the Kansas telescope is about 17 degrees. In contrast, the 250-foot reflector of the radio telescope at Jodrell Bank in England resolves the sun as an object about twice the diameter of the optical disk. Toy spyglasses can disclose much more detail. But resolving power is only one measure of a telescope's performance. Another is the instrument's ability to detect distant objects. The clouds of interstellar dust that block many regions of the universe from view are transparent to some bands of the radio spectrum. The amateurs who built the Winston-Salem and Kansas telescopes set out to have a firsthand "look" at whatever lies beyond the dust, even if the view turned out to be fuzzy.

"I began to work on my telescope," writes McFarland, "during my third year in college, partly as a project for a thesis, and I hoped to finish it before graduation. But a number of bugs developed, and it was not ready for a trial run until the summer following graduation. The telescope has four major components: the antenna and its mount, a high-gain, low-noise receiver, an automatic pen recorder and a noise generator that is used to test the system and as a standard for comparing the strength of radio sources in space. [These days the pen recorder is usually replaced by a personal computer and an analog to digital (A to D) converter. Ed] The design and procurement phase of the project took 18 months of spare time and the construction about a year.

"Much of the initial planning went into the antenna. The antenna of a radio telescope corresponds to the objective mirror or lens of an optical telescope, and the performance of the completed instrument depends on it just as critically. In selecting a design for the antenna several configurations of the diffraction type were considered, including a broadside array of half-wave dipoles and an array of helices. These were dismissed in favor of a paraboloid because the complexity of interconnecting a broadside array increases in proportion to the number of dipoles, and the length of each dipole must be changed for each frequency on which the telescope operates. Moreover, I wanted an antenna that would pick up the broadest possible band of frequencies and discriminate strongly against all signals except those that come from a desired direction. A paraboloid best meets these requirements.

"Winston-Salem is a center of intense, man-made electrical disturbance, chiefly from sources such as automobile ignition systems, power lines and harmonic radiation from radio and television stations. By scanning the radio spectrum from 50 to 3,000 megacycles with a short-wave receiver, I spotted a relatively quiet region of the spectrum in the vicinity of 400 megacycles (a wavelength of 75 centimeters, or 29½ inches). At this frequency a signal equal to a millionth of a billionth of a watt (10^{-15} watt) would override the noise if the antenna were designed for maximum power gain; that is, if it strongly favored signals arriving parallel to the axis of the parabola. The power gain of a paraboloidal antenna (with respect to the response of a nondirectional antenna) varies directly with the radius of the parabola and inversely with the wavelength of the signal, as shown by the accompanying equation:

Power Gain of Paraboloid
Antenna Relative to a Simple
Dipole Antenna

$$\text{Gain}_{max.} = 10 \log_{10} \left(\frac{2\pi}{\lambda} \frac{2fr^2}{4f^2 + r^2} \right)^2$$

where:

f = focal length of paraboloid
r = radius of paraboloid
λ = wavelength of signal

Power-gain equations

When the focal length of the parabola is equal to half the radius, the maximum power gain in decibels is equal to 10 times the logarithm (to the base 10) of the square of this ratio: 3.14 times the radius divided by the wavelength. At a wavelength of 27.5 inches (a frequency of approximately 436 megacycles) and a gain of 20 decibels, this formula yields a radius of 7.5 feet. With this dimension known, the distance from the focus to the vertex of a paraboloid can be calculated. In the case of my antenna it amounts to 3.75 feet. The resolving power of telescope objectives, whether optical or radio, increases in proportion to the diameter of the lens or reflector, and decreases with wavelength as indicated by the second formula [*see page 75*]. A 15-foot paraboloid operating at 436 megacycles has a resolving power of 10 degrees 54 minutes, which is about 20 times greater than the apparent angle subtended by the sun.

"With the size of the antenna determined, its physical structure was considered next. Aluminum was selected as the most attractive material, from the point of view of both weight and cost. A disadvantage in using aluminum is that all parts of the antenna must be welded. Otherwise voltage may develop across the high-resistance joints between adjacent parts and be detected as noise. The welding can be done most satisfactorily by

Resolving Power of Telescope Objectives

$$\theta = 1.22 \frac{\lambda}{a}$$

where:

θ = resolving power in radians

λ = wavelength

a = aperture

(λ and a in same units of length)

For White Light

$$\theta = \frac{14.1}{a}$$

where:

a is in centimeters

θ is in seconds of arc

Resolving power of 508-centimeter (200-inch) aperture
for white light would be:

$$\theta = \frac{14.1}{508} = 0.02776 \text{ seconds of arc}$$

For radio waves of 300 mega-cycles (100 centimeters) the
resolution of a 508-centimeter paraboloidal reflector would be:

$$\theta = \frac{1.22 \times 100}{508} = 0.24 \text{ radians} = 13° \ 45'$$

and for 436 megacycles (69 centimeters) the resolution of a 15-foot paraboloid
(457 centimeters) would

$$\text{be: } \theta = \frac{1.22 \times 69}{457} = 0.193 \text{ radians}$$

Resolving-power equations

an electric arc that operates in an atmosphere of helium gas. This is inconvenient but not expensive.

"The paraboloid was formed of sheet aluminum welded to a paraboloidal skeleton of aluminum tubing—a series of concentric rings supported by radial ribs bent so the sheet took the desired shape to within ⅛ inch [*see illustration on page 76*]. To build the skeleton I formed nine circles of ¾-inch tubing, with radii ranging from seven feet nine inches to six

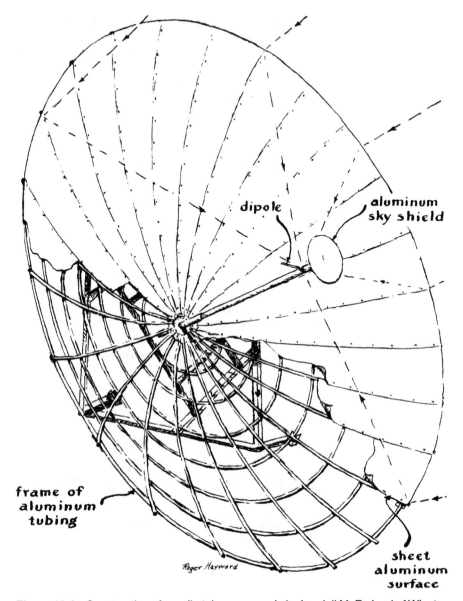

Figure 10.1 Construction of a radio telescope made by Lyndall McFarland of Winston-Salem, N.C.

inches, on a machine similar to those used for bending model railroad tracks. The rings were nested against 24 radial ribs formed to the approximate final shape on the same bending machine. Many hours were then spent in hand-forming each rib to within ¹⁄₆₄ inch of a master parabolic template cut from sheet aluminum. A jig clamped the parts during the welding operation.

"Aluminum screening would doubtless have been a better choice from the point of view of wind resistance for covering the skeleton. But the only available material of this sort was ordinary house screening, which is much too light to hold its shape or to weld to the skeleton. The completed structure weighs approximately 260 pounds.

"Incoming signals are focused on a dipole antenna that is supported on the axis of the paraboloid by a short length of coaxial transmission line made of two aluminum pipes [*see illustration on page 78*]. The electrical impedance of coaxial lines is equal to the logarithm (base 10) of the quotient of 138 times the inside diameter of the outer pipe (in inches) divided by the outside diameter of the inner pipe. The inside diameter of my outer pipe is .66 inch and the outside diameter of the inner one is .27 inch. The impedance of my coaxial line is therefore 53 ohms. For maximum transmission efficiency the impedance of the coaxial line and that of its associated dipole antenna must match. The characteristic impedance of a dipole antenna in free space is 72 ohms, but this value is lowered by the presence of a nearby conductor, such as a metal plate. I found that I could match the impedance of my dipole to that of the coaxial line by placing an aluminum disk (16 inches in diameter) a quarter of a wavelength in front of the dipole. Later I found that the disk also helped to shield the dipole from off-axis signals and therefore improved the directivity of the antenna.

"The coaxial line is approximately five feet long and is fastened to the paraboloid through an aluminum plate welded to the skeleton. The outer end is terminated in threaded fittings that take a pair of threaded aluminum rods, each a quarter of a wavelength long, which function as the dipole. The outer pipe of the coaxial line extends approximately two feet beyond the dipole and serves as a mounting for the aluminum disk. The inner end of the coaxial line is equipped with a threaded coaxial coupling for attaching the antenna to the receiver through flexible coaxial cable.

"Sky noise from the cable is fed at 436 megacycles to a parametric amplifier and gains some 20 decibels in power. The signal is then passed to a conventional converter that divides the frequency to 30 megacycles. After additional amplification the signal is converted to pulsating direct current to run a pen recorder. All the apparatus is powered from closely regulated power supplies.

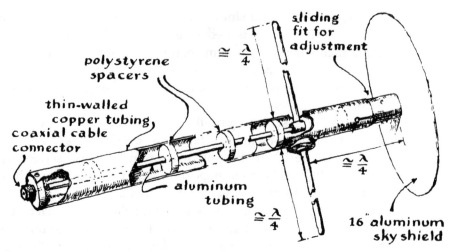

Figure 10.2   Coaxial feed line of McFarland's antenna

"During the initial tests and sightings the paraboloid was mounted on a meridian transit, but as soon as time permits it will be installed on a surplus 36-inch searchlight mount equipped with a motor drive for remote control.

"The parametric amplifier turned out to have substantially more gain than expected, so I was able to use a long section of coaxial cable between it and the converter without introducing excessive noise. The gain was so great, in fact, that I could even install a six-decibel attenuator in the cable to keep the ignition noise of passing cars from driving the converter into overload.

"There has not been time to use the telescope extensively since it has been completed. But test runs prove that its response is satisfactory, considering the comparatively low resolution of the system. One sad incident is worthy of mention because it disclosed that parabolic antennas of sheet aluminum must be painted flat black. While I was observing the sun during the first trial run the 16-inch aluminum disk at the focus of the paraboloid suddenly melted!"

The antenna of the Kansas telescope, according to Houston, Simpson and Mullinix, was constructed primarily for tracking satellites. It consists of a 35-foot spar of pipe that supports a single reflector and a dipole antenna at the back and a series of dipole directors in front. The spar is supported by two braces of pipe and carries an altitude circle [*see illustration on page 79*]. The directors and reflector are merely straight lengths of aluminum wire ⅛ inch thick. The wires stand up well under the Kansas winds, according to Houston, but birds can bend them. The whole business is mounted on an 18-foot telephone pole so that the reflector just clears the ground when the antenna points to the zenith.

"For a given power gain," Houston writes, "Yagi antennas can be built that are lighter and more compact than any other type and that have less wind resistance. These advantages are bought at the cost of a narrow band width. Yagi antennas can be designed for optimum operation at only one frequency—a real disadvantage when they are used for measuring star noise.

"When in operation, the antenna is fixed at a selected altitude on the meridian for 24 hours and picks up noise in varying amounts as the sky drifts past. The sky noise is amplified and converted to pulsating direct current for operating a pen recorder. [Again, replace the pen recorder with an A to D converter. Ed.] Normally the antenna is pointed just above the horizon during the first run of a series and then is raised a few degrees higher for each subsequent run until the entire sky has been scanned from the horizon to the zenith. If the recorded traces turn out to be good, corrections are made for instrumental errors and the results are read off and plotted.

"The readings are somewhat fictitious, of course. Just as an optical telescope shows spurious disks around the stars, so do radio telescopes. By running the sun across the antenna we found that the disk of our instrument is about 17 angular degrees in a horizontal direction. We did not measure its vertical width but it should be about the same. 'Radio stars' and other discrete sources of radio sig-

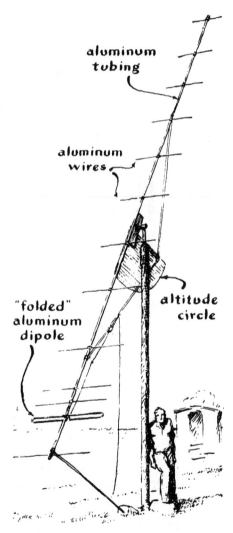

Figure 10.3 Yagi antenna of the Kansas telescope

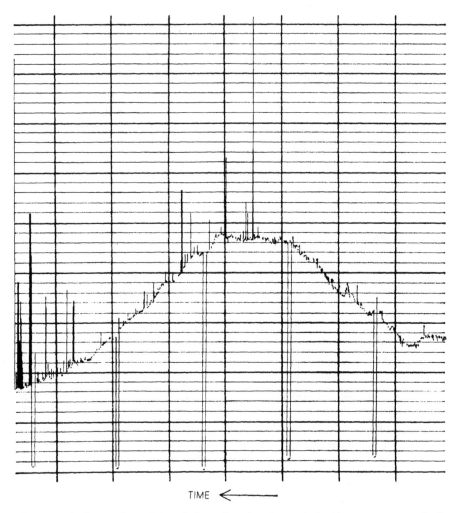

Figure 10.4 Recording of signals from Cassiopeia made by Kansas amateurs' telescope

nals are below the resolution of the system and are lost in the general background noise.

"Our telescope consists of the antenna, a preamplifier, a frequency converter, a wide-band intermediate amplifier, a detector and a recorder. The sensitivity of the system is limited by the noise generated in the preamplifier. (Disturbances from artificial sources are not serious in our locality.)

"The output of the frequency converter is fed directly to the first intermediate amplifier stage of an old television receiver. The band width of a correctly aligned television receiver is about five or six megacycles, whereas that of our Yagi antenna is on the order of two to four megacycles. Reducing the band width of a radio telescope is comparable to filtering the reds and blues from the ends of the optical spectrum: the energy of the signal is reduced, and in the case of radio telescopes the maximum amount of sky noise is wanted. The weak link in our telescope, however, is the narrow band width of the antenna, so the television amplifier is adequate.

"The output of the television amplifier is rectified by a diode and, after passing through a resistor and into a capacitor, it is fed to the pen recorder. The time required for the capacitor to charge is about a tenth of a second. This delay tends to smooth the recorded graph because the pen does not respond to current pulses of less than a tenth of a second, such as bursts of lightning.

"Our equipment—most of it salvaged from the scrap box—can only be characterized as crude. It operates on 108 megacycles, far below the 1,421-megacycle hydrogen line that is so widely observed by radio astronomers. But, like most amateurs, we have a fondness for making our initial forays with the help of salvaged junk and in areas that are neglected by the hunters of bigger game.

"We set out to make a radio map of the sky—and we made one that agrees broadly with those compiled by the best radio telescopes, considering differences in frequency and antenna resolution [*see illustration on page 82*]. In plotting the map we made no effort to compute away the influences of discrete radio sources in space. Although these powerful cosmic radiators do not show up as spots on our map, they do tend, like an overexposed star on a photographic plate, to distort our contour lines. This is particularly apparent in the lines that dip to the south from Cassiopeia. Even more striking is the influence of the two radio stars in Cygnus near 40 degrees declination and 20 hours right ascension. Another major difference is the hourglass configuration in Orion. One map made by a large radio telescope at 250 megacycles shows two local areas in Orion, one at the Equator around the 'belt' and one 10 degrees north and somewhat toward the east. The pinch in the hourglass of this map falls at about 10 degrees south declination, whereas ours lies directly on the Equator. This difference worried us, and we made a dozen extra runs in the region of Orion, even making runs in declination by hand. But our results were always the same. We now suspect that the appearance of this region of the sky may be different at 108 megacycles from its appearance at 250 megacycles.

"In spite of the fact that we ran traces at each declination for several days, and in some cases for many days, we never succeeded in recording

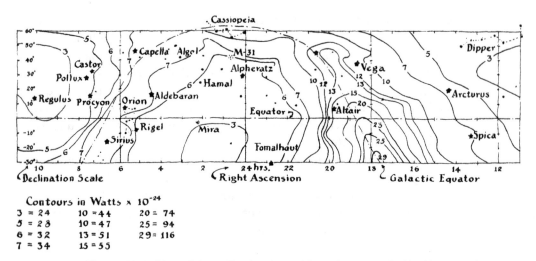

Figure 10.5 Map of the radio sky charted from data recorded by Kansas telescope

a full 24-hour run at any one time. The sun was usually evident, and if it wasn't, we did not trust that portion of the curve. To map such regions of the sky we simply had to wait until the sun moved on. On other occasions the electronic equipment went psychopathic and awed us by recording giant forces that never reappeared; now and then thunderstorms ran the pen out of ink. In general we got one good trace out of 10 tries. But it was fun."

Amateur radio astronomy has come a long way in the years since this article appeared in *Scientific American*. Back in 1962, few amateurs had built radio telescopes and there was no organized effort devoted to promoting amateur radio astronomy. Fortunately, that's all changed. Hundreds of amateurs can now observe cosmic radio waves using their own home-brew equipment. Here are three Web resources to help you get started.

Society of Amateur Radio Astronomers (SARA)

This groups helps amateurs advance their interest in amateur astronomy. You can visit their Web site at

www.bambi.net/sara.html

Project Bambi

This world-class radio telescope was built entirely by dedicated amateurs. For inspirational details, set your browser to

www.bambi.com

RadioSky Publishing

This company provides access to the best information available on radio astronomy. To find out what's available, check out

www.radiosky.com

Finally, the search for extraterrestrial intelligence (SETI) is being carried out entirely by radio astronomy. In fact, the professionals have collected far more data than they can analyze. So a visionary team at the U.C. Berkeley's Space Sciences Laboratory has created a way for you to do just that. So surf over to setiathome.ssl.berkeley.edu and download *SETI@Home*. This innovative program will run in the background and direct any spare CPU time on your personal computer to look for ET. The interface is well-designed and it's quite interesting to watch the program run. My copy runs whenever my computer is free. I can't think of a better way for my computer to spend its spare time.

Ed.

PART 2

THE SUN

11 HOW TO OBSERVE AND RECORD SUNSPOTS SAFELY

By Forrest M. Mims III, June 1990

To the casual observer, the sun is immutable. It shines constantly with a bright white light that changes to reds and yellows only when scattered or absorbed by particles and vapors in our planet's variable atmosphere. A closer look reveals, however, that the sun is far more dynamic than the earth. Cataclysmic storms periodically erupt on the sun's surface—storms that could easily envelop several earth-size planets.

Regions of intense activity on the sun's surface are somewhat cooler than the area that surrounds them. For this reason, active regions appear dark when viewed against the hotter and therefore more brilliant solar disk. These active, dark regions are known as sunspots.

Some spots cover such a large fraction of the solar disk that they can be seen without magnification. Indeed, more than 1,700 years before the invention of the telescope, Chinese astronomers observed sunspots without the assistance of a magnifying device. Although I have long been aware of such ancient reports, not until March, 1989, did I personally look

> **CAUTION:** Viewing the sun with the naked eye or through any kind of magnifying instrument can result in severe damage to the eyes, including permanent blindness. Before you perform any of the experiments described here, you should follow the given instructions carefully. Children should not attempt these observations without adult supervision.

at sunspots without a special telescope. In separate accounts on the same day, two acquaintances informed me that they had seen a large spot on the sun while driving to work. The sun was low on the horizon, and they were able to view it safely through a layer of fog.

By the time I received these reports, the fog had long since vanished. I therefore drove to a nearby welding store and purchased a filter plate that is designed for an arc-welder's helmet. Within a few minutes the store's staff and I were outside looking at a large group of sunspots through plates of welder's glass.

Serious sun watchers will recall this unusual group of some 50 spots by its formal name, region 5395. Thanks to widespread media coverage, the rest of us will remember region 5395 as the cluster of spots that gave rise to a solar flare, which in turn caused a spectacular, luminous display in the night sky over most of the Northern Hemisphere. This display, known as an aurora borealis, appeared as far south as the Florida Keys and Cancun, Mexico.

When an aurora appears that far south, you can be sure that someone, somewhere, wishes it had not. That is because the increased solar activity that generates the aurora can have many deleterious effects on and near the earth. The orbits of satellites, particularly those in low orbit, can be altered; radio communications can be disrupted; and electric power grids can be subjected to power swings and even blackouts.

Region 5395 caused more than its fair share of such mischief, a compilation of which was prepared by Joe H. Allen of the World Data Center-A for Solar-Terrestrial Physics in Boulder, Colo. Power fades or outages occurred in New Mexico and New York. Six million residents of Quebec Province went without electricity for nine hours or more on March 13. These blackouts cost power utilities a total of 187 million kilowatt-hours.

Sunspot region 5395 was also responsible for many problems with broadcast systems. The earth's upper atmosphere usually refracts radio signals, but it absorbs them when it is bombarded by intense solar radiation. The opposite is true for higher-frequency signals: they propagate far beyond their usual range. During the lifetime of region 5395, these effects provided many interesting experiences for both amateur and professional operators of radio-communications systems. The effects also explain why some homeowners in California complained that their radio-controlled garage doors were mysteriously opening and closing on their own. That phenomenon was apparently caused by a nearby Navy transmitter. The station had shifted operation to a new frequency, because its standard frequencies were rendered useless by the effects of the sun on the ionosphere.

Despite all the trouble that sunspots cause, they can be a pleasure to watch. And you can gain a better appreciation of the dynamic nature of the sun by observing it daily for a month or so. You will find this project informative and considerably more convenient than those conducted by the professional and amateur astronomers who study stars other than the sun. Whereas these investigators often stay up all night in pursuit of their quarry, you will be able to observe the sun anytime during the day. You can even watch through haze or smog or from any room that has a window facing the sun.

The simplest and fastest way to look for large spots is to look at the sun directly through a suitable filter, as I first did during the appearance of region 5395. It is absolutely essential that the filter attenuate the sunlight to the proper degree. Welder's filters are rated according to the amount of light they transmit. The darkest available filter has a rating of 14. It transmits 2.7 times less light than a number 13 filter, which transmits 2.7 times less light than a number 12 filter and so on. Only a number 14 filter provides enough protection for your eyes against direct sunlight.

Welder's filters are available in at least two sizes: five by 10.8 centimeters and 11.4 by 13.3 centimeters. The small filter provides a pocket-size solar observatory that you can carry with you to check for large spots during traffic jams, lunch breaks and hikes. The large filter, which will fit in a coat pocket or purse, provides somewhat more comfortable viewing since it shades most of the face. Since some welding stores do not stock filters with a shade darker than number 12, it is a good idea to call first.

Some welding stores stock plastic filters coated with a metallic film. Although these filters have the same attenuation factor as glass filters marked with the equivalent shade number, a scratch in the metallic coating can allow the transmission of damaging rays. For this reason, a glass filter is a better choice.

No matter what kind of welder's filter you purchase, under no circumstances should you attempt to use such a filter in conjunction with a telescope or binoculars! These instruments gather more than enough light to damage your eyes, even if they are protected by a filter. [Actually, welder's glass no. 14 will allow you to view the sun safely if it is firmly duct taped over the front of both binoculars. But be warned. If the filter should become loose or fall off, you are risking your eyesight! Ed.] Furthermore, a glass absorption filter placed between your eye and a telescope's eyepiece can be shattered by the intense heat caused by the magnified image of the solar disk. [Absolutely right. I know one person this happened to. If you place the filter over the eye pieces you are courting disaster. Ed.]

The sun appears yellow or yellowish-green through a glass welder's filter and gold through a plastic filter. The glass filter can be tilted to increase its attenuation, a helpful adjustment when attempting to discern small sunspots. If your first glance discloses no obvious spots, look at the edge of the sun and then the entire disk. If visible spots are present, one or more may pop into view when you shift your view back toward the full disk. Indeed, that is precisely what happened when I stepped outside in the course of typing this paragraph. A first look revealed a clear solar disk. After shifting my field of view around several times, suddenly two large spots appeared on opposite sides of the sun's equator.

A large welder's filter will let you try a viewing-enhancement trick I serendipitously discovered. While watching the sun through the filter, tilt the top side of the filter away from your forehead until a blue patch of sky is reflected into your eyes. A blue field will now be superposed over and around the solar disk. Because the filter is tilted, the sun will appear dimmer. The reflected skylight will contract the pupils of your eyes, thereby making the sun appear dimmer still. In my experience this technique greatly enhances the visibility of sunspots.

This rather primitive observation method can be more rewarding than it might at first seem. You might have the opportunity to watch the progress of a major sunspot group. In late August of 1989, for example, my pocket-size observatory revealed a very large spot on the sun's east limb. Three days later the spot had grown into two giant spots connected to each other—a fact that confused even professional astronomers. At first the two spots were assigned separate names, regions 5669 and 5671. When astronomers analyzed the magnetic structure of these two regions, however, they found the spots were indeed a single massive sunspot group. So regions 5669 and 5671 were combined and designated region 5669.

This highly unusual sunspot group was the source of several major bursts of radio waves, light waves and X rays. I followed region 5669 for nearly two weeks until it rotated over the sun's west limb. One morning a few minutes after sunrise, I had a rare opportunity to see region 5669 through a thick haze and ordinary sunglasses.

Region 5669 was so large and its geometry so unusual that I wanted to examine its structure in greater detail. I therefore used a pair of binoculars to project a clear image of the solar disk onto a sheet of paper. The easiest way to implement this method is to mount the binoculars on a camera tripod. Some camera stores sell an adapter that will let you mount many kinds of binoculars on a tripod.

Under no circumstances should you look through the binoculars while attempting to aim them at the sun! Nor should you sight along the side or top of the binoculars. Instead, place a lens cap over one of the two apertures, and point the binoculars in the approximate direction of the sun. Then adjust their position while watching their shadow. When the shadow has the smallest profile, the binoculars are almost properly aligned. At that point, move them slightly until a faint image of the solar disk appears in the shadow. Align the binoculars until the disk is nearly centered in the shadow. When it is, place a cardboard light shield over the front of the open lens in order to shade the image of the solar disk and make it much brighter. Then place a white sheet or card 20 to 30 centimeters away from the binoculars, and carefully focus the binoculars for the sharpest image.

The projection method of viewing sunspots can be implemented with most binoculars and telescopes and is by far the safest method. It is important, however, to realize that curious children might attempt to look through the eyepiece of an instrument pointed at the sun. You must therefore supervise children (and adults who should know better) who are near any optical instrument that has been aimed at the sun. For even the briefest glimpse at the solar disk through a small telescope can cause a permanent and significant loss of vision. You should also be aware that binoculars or telescopes fitted with a reticle can be damaged if they are used to project solar images. The intense, focused light from the sun can burn cross hairs and melt plastic reticles.

If you have a small telescope, you can easily assemble a permanent solar-projection observatory. Two years ago Vicki Rae Mims, my teenage daughter, did just that. Vicki constructed her system from scrap lumber, a clipboard, a cardboard shade and a small 10-power telescope. A finder telescope from a larger telescope will work fine. You may also purchase a small telescope from a supply company, such as Edmund Scientific Company (101 E. Gloucester Pike, Barrington, NJ 08007, *www.edmundscientific.com*). Vicki's observatory is illustrated on page 92. For your own solar observatory, you might want to increase the distance between the telescope and the clipboard in order to enlarge the projected solar disk.

To operate your observatory, you should lean the telescope end of the device against a fence or on one of the rungs of a stepladder. Remove the cardboard shade from the telescope, and move the ground end of the observatory toward or away from the support until the telescope's shadow is centered on a sheet of paper held fast by the clipboard. The telescope will now be pointed approximately at the sun. You can then make fine adjustments until the solar disk is centered on the sheet of paper. Then replace the cardboard shade to brighten the image of the sun.

With this simple projection system, you can track the movement of spots across the solar disk. You will first need to center the disk on a sheet of paper held by the clipboard. Mark several points around the perimeter of the projected image, and then draw a circle through the points. You may wish to add a system of coordinates or a grid pattern so you can better specify the location of interesting spots. Provide places for the date, time and comments. Make this sheet of paper your master chart, and copy it.

To record the movement of sunspots accurately, you should make your observations at the same time each day. This schedule guarantees the sun

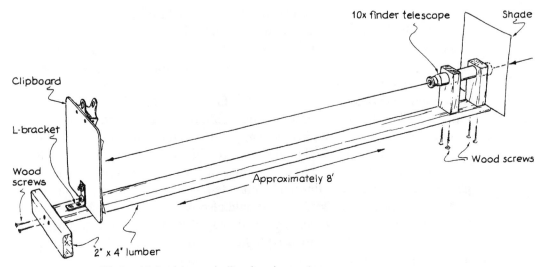

Figure 11.1 A home-built solar observatory

will be oriented in the same way each time you observe it. If your daily schedule will not permit this, align the system, and watch for a few minutes as the sun's image drifts across the chart. If necessary, rotate the clipboard or the entire observatory so that a sunspot moves along or parallel to a previously drawn grid line.

Eventually you should find the approximate north pole on the solar disk. At local apparent noon in the Northern Hemisphere, celestial north is at the top of the sun. The actual north pole will be within 26 degrees of celestial north. If a scene viewed through your telescope's eyepiece is inverted, the top of the projected image of the sun at noon is north. If not, the top of the projected image is south.

For four months Vicki made almost daily observations of sunspots using this system. She measured the rotation of the sun by tracking several spots and groups completely across the solar disk. Vicki observed that some spots moved across the solar disk faster than others, a difference that occurs because the gases at the solar equator rotate more rapidly than the gases toward the poles.

While observing projected sunspots, periodically move the paper back and forth. This technique will help remove the effects of the paper's surface texture and bring out details you might have missed. Although the images you see can be saved with the help of a camera or video recorder, many amateurs prefer to mark the sunspots with a pencil, as Vicki did. Doing so lets you indicate small spots that might be missed by film or a video camera. But, unless you have an automatically guided telescope, you have to work fast. If there are many spots, you will have to realign the telescope periodically to make sure the solar disk stays superposed over the outline on the paper.

Serious watchers may prefer to monitor the sun with an astronomical telescope that allows either projection or direct viewing. Direct-viewing instruments are usually equipped with metallic-film aperture (not eyepiece) filters. An instrument of reasonable quality will reveal that the central dark portion, or umbra, of some sunspots is surrounded by a lighter region known as the penumbra. You may also be able to classify sunspots and sunspot groups according to their size and appearance. One classification system is described in the illustration on page 94. You may wish to consult the references that follow for additional information and safety precautions.

Whether or not you elect to pursue a regular program of sunspot obser-vations, you will find many other ways to keep up with the latest develop-ments on the sun. A brief summary of current solar conditions and related geophysical activity is broadcast at 18 minutes past each hour by radio sta-tion WWV. The message is updated every three hours. You can receive WWV on a shortwave receiver at frequencies of 2.5, 5, 10, 15 and 20 MHz. Because WWV continually broadcasts precision-time measurements, you can set your watch while waiting for the latest solar-activity update. If you do not have access to a shortwave receiver, the WWV message is also avail-able by telephone. The number is (303) 497-3235.

The Space Environment Services Center in Boulder operates an experimental Public Bulletin Board System (PBBS), which will send cur-rent information and forecasts about solar activity to a personal com-puter equipped with a modem. You can access the PBBS by dialing (303) 497-5000. (To access the system, you will need to know that the proto-col is an eight-bit data word with one stop bit and no parity at either 300 or 1,200 baud.) The web site is *www.neonet.nl/ceos-idn/datacenters/ NOAAOARERLSECSWO.html.* For more solar web sites, see page 98.

Another excellent way to keep up with solar events is to subscribe to the *Preliminary Report and Forecast of Solar-Geophysical Activity.* This weekly publication gives highlights of solar and geomagnetic activity and forecasts activity for the next 27 days. The official sunspot number is reported, and a complete list of X-ray and optical flares is given. Of par-ticular interest are the graphs comparing the current solar cycle with pre-vious cycles. If you cannot find the report at a nearby technical library, you can purchase a subscription from the Space Environment Services Center, NOAA R/E/SE2, 325 Broadway, Boulder, CO 80303-3328. See the next chapter for a list of interesting solar web sites.

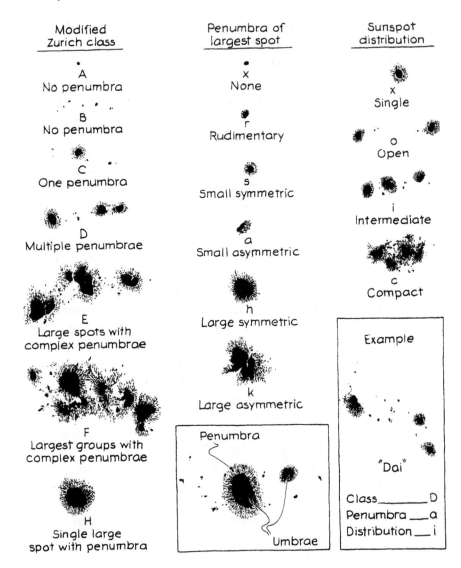

Figure 11.2 The McIntosh system of sunspot-group classification

Further Reading

(Readers may need a book dealer to find some of these older texts)

A Complete Manual of Amateur Astronomy. P. Clay Sherrod with Thomas L. Koed, Prentice Hall, 1981.

Observational Astronomy for Amateurs. J. B. Sidgwick. Dover Publications, Inc., 1981.

The Sun. Iain Nicolson. Rand McNally and Co., 1982.

Watching the Premier Star. Patrick S. McIntosh and Harold Leinbach in *Sky & Telescope*, Vol. 76, No. 5, pages 468–471; November, 1988.

12 SUN OF A GUN

By Shawn Carlson, August 1999

I won my first telescope when I was nine years old by selling 500 boxes of flower seeds door-to-door. (I was a good talker even then.) It wasn't much of an instrument, just a four-inch refractor that suffered from what astronomers call chromatic aberration: it focused different colors at slightly different distances, so that only one color could be in focus at a time. Stars and planets were so blurred that I almost relegated the telescope to my closet. But it was saved by its sun filter, which allowed a smidgen of the sun's green light to pass through. I gasped out loud the first time I used it. Limited to just one color, the solar disk came in razor-sharp, and sunspots appeared like large black islands in a vast emerald sea.

That experience inspired my first amateur research project. Every day that summer at precisely 11:00 A.M., I set up my telescope and carefully sketched the sunspots on a piece of graph paper. I quickly discovered that the sun's surface, unlike the earth's, rotates at different rates depending on latitude. Sadly, my intensive investigations soon wore out the scrawny scope. Since then, I've visited our home star mostly through no. 14 welder's glass duct-taped over binoculars and recently via the World Wide Web [*see box on page 98*].

But the total eclipse that cut across Europe and western Asia on August 11, 1999 put me on the lookout for better ways to see the sun. So you can imagine my excitement when I learned of an elegant solar projector designed by Bruce Hegerberg of Norcross, Ga. It creates a dazzling daylight display. The so-called limb-darkening effect—that is, the apparent drop in brightness near the sun's edge caused by the longer viewing path through the sun's atmosphere there—is plainly visible. Also, the characteristic structure of sunspots, with a dark inner umbra surrounded by a lighter penumbra, can be clearly seen.

Because the solar image can be easily viewed in daylight by many people at once, Hegerberg's fabulous device is perfect for eclipse watching.

Moreover, by presenting such enticing images during the day when it is easiest to reach nonastronomers, this projector could revolutionize sidewalk astronomy—the time-honored practice whereby amateur astronomers set up small telescopes to give passersby a peek at the heavens.

Hegerberg fashioned his first solar projector, the "Sun Gun," from an inexpensive telescope assembly, some PVC piping and a large flowerpot. Those interested in the details should check out his Web site. Here I will describe his second-generation device, the "Sun of a Gun," which can be quickly and cheaply assembled from a paint bucket. If your telescope has a heliostatic (sun-following) motor drive, you'll be able to track the sun's motion for hands-free viewing.

You'll need a plastic five-gallon (20-liter) paint bucket (such as Home Depot part no. 084305355553). Discard the lid and paint the inside of the bucket black to prevent ambient light from coming through the translucent plastic. Cut a 2¼-inch hole in the bottom using a hole saw attached to an electric hand drill. Through the hole, thread a male flexible adapter for a water hose (Ace Hardware part no. 45708) and secure it in place with one two-inch conduit locknut (Home Depot part no. 051411461966). (Obviously, readers outside the U.S. will need to adapt these measurements to a metric equivalent, depending on the availability of hardware.)

Next, drill an ⅛-inch hole about a half-inch from the end of the adapter. Line up this hole with the screw hole in the eyepiece assembly and lock the two together using the screw that normally holds the eyepiece in place. If the adapter does not fit your scope, affix a universal camera adapter (about $30 from Orion Telescopes; 800-676-1343 or *www.telescope.com*) to your scope and attach the bucket to that.

The sun's image appears on a rear-projection screen of the kind often used in large-screen TVs. Many varieties of screen are available, each with different trade-offs in viewing angle, image brightness, sharpness and contrast. Hegerberg purchases a flexible Da-Tex rear-projection screen for $10 per square foot from Da-Lite Screen Company (800-622-3737 or *www.da-lite.com*). A 15-inch square will suffice. Secure the screen, polished side facing out, over the open end of the bucket. You can use a 48-inch plastic wire tie positioned just under the bucket's lip. The tie is the same type that can bind large bundles of wire, and Home Depot has them (part no. 728494104805). Pull the screen taut as you tighten the tie, so that the assembly resembles a drum. Alternatively, you can secure the screen with a large rubber band. Cut off the excess screen, leaving about a half-inch of fabric below the tie for future adjustments.

Finally, Hegerberg removes the bucket's handle and slips a large rubber band over it. After reattaching the handle, he connects the band to the finder scope to relieve some of the stress on the focusing assembly [see *illustration below*]. Depending on the size of your bucket and scope, you might also need to add a counterweight to the telescope tube.

To get a clear image of the sun, you'll need a good eyepiece and a filter that screws into it. Hegerberg recommends Plössl eyepieces because they deliver the sharpest and best color-corrected images, but Huygenian eyepieces contain no cemented elements and so may better survive long-term exposure to the sun's heat. You'll need focal lengths between 17 and 25 millimeters depending on the size of your telescope. If you happen to own a Schmidt-Cassegrain telescope, try a 20-millimeter eyepiece for a four-inch instrument and a 25-millimeter eyepiece for an eight-incher. Sirius Plössl eyepieces retail for about $50 from Orion. If your telescope's aperture is larger than four inches, you must attenuate the light using a piece of cardboard with a four-inch hole in it. Attach this cardboard to the front of your scope. Otherwise, your instrument could overheat.

For the filter, Hegerberg recommends #21 (orange), #11 (yellow green) and #12 (yellow), any of which Orion sells for about $15. But keep in mind

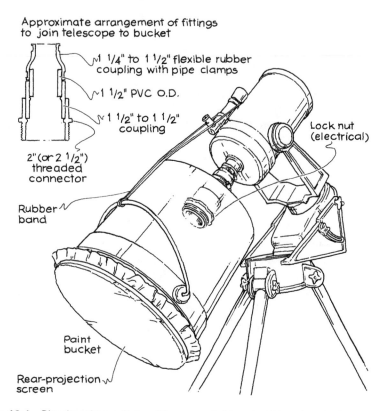

Figure 12.1 Simple solar projector diffuses sunlight so that it is safe to look at. It attaches to the eyepiece of an ordinary amateur telescope.

Solar Web Sites

The National Aeronautics and Space Administration's guide to the August 11 eclipse includes detailed maps of the path. *sunearth.gsfc.nasa.gov/eclipse/TSE1999/TSE1999.html*

NASA's Solar and Heliospheric Observatory Web site has real-time solar images and movies of recent solar activity. *sohowww.nascom.nasa.gov*

The National Solar Observatory's site shows the sun in various wavelengths. *www.nso.noao.edu/synoptic*

NASA's Solar Data Analysis Center has a collection of archived solar images. *umbra.gsfc.nasa.gov/images*

The Space Environment Center offers current and archived images. *www.sel.noaa.gov/solar_images/ImageIndex.cgi*

The Association of Lunar and Planetary Observers (ALPO, *www.lpl.arizona.edu/alpo/*) runs a site full of amateur and professional solar images. *www.lpl.arizona.edu/~rhill/alpo/solstuff/recobs.html*

Bruce Hegerberg's site offers more information on the project in the article. *www.america.net/~boo/html/sun_gun.html*

Ed.

that these filters were never intended for direct solar viewing. Just as you would never press your eye over the lens of a movie projector, so you should never look directly into the eyepiece—even with one of these filters. Doing so could permanently damage your vision. The projection screen on the Sun of a Gun diffuses the light so that it is safe to look at.

Because the finder scope can focus sunlight enough to cause burns, always cover it before using the Sun of a Gun. Of course, never look through the finder scope at the sun. To align the telescope with the sun, first adjust its position so that it casts the smallest possible shadow. Then use the focus to sharpen the image on the screen.

Armed with this powerful tool, you'll be ready to explore our home star on any clear day. You, too, may enjoy observing the life cycle of sunspots, recording the ratio of the umbra to penumbra area or mapping their size over time.

13 A CORONAGRAPH TO VIEW SOLAR PROMINENCES

Conducted by C. L. Stong, September 1955

Except for two obstacles you could raise your thumb to the sky on a clear day, line up the nail with the edge of the sun and see tongues of blood-red flame lashing thousands of miles into space. You are cheated out of this exciting spectacle by your heterochromatic eyes and the dense, dirty atmosphere of the earth. Glare refracted by dust, water vapor and the molecules of the air masks the relatively faint rays of these solar prominences. The problem of suppressing the glare and sorting the red rays from the white residue constitutes one of the most interesting challenges to the amateur astronomer.

Prominences show up clearly during a total eclipse of the sun, when the glare is masked by the moon. Why not equip a telescope with an artificial moon—an opaque disk just large enough to blot out the sun? Such a disk would be merely an elaboration of your thumbnail, and it would fail for the same reason that the thumbnail fails. The disk of the real moon is located in airless space: hence it casts a knife-sharp shadow. Your artificial moon would cut off the direct rays of the sun but could not stop the rays reflected from all angles by the dust-laden atmosphere. You must equip your telescope not only with an artificial moon but also with a filter which can distinguish between direct rays from the disk of the sun and those from the prominences. This requirement implies a difference between the two kinds of light, because you cannot sort things which are identical.

It turns out that prominences are largely composed of hydrogen, which emits a deep red light. This light is emitted in discrete bands, corresponding to the energies liberated when the electrons fall from excited atomic states to those with lower energy. To rid the system of the electron's excess orbital energy, the system creates a photon with an energy precisely equal to energy difference between the two states. Most of the hydrogen light from the sun comes about from the so-called Hydrogen-alpha transition which occurs when an electron from an excited state falls to the ground state. These photons emerge from the sun inside a band just 1.2 angstroms wide and centered at a wavelength of 6562.81 angstroms. The sun as a whole, on the other hand, is composed of all elements, each of which contributes one or more colors to the visible spectrum. White light is a mixture of all these colors. Thus the problem of seeing prominences comes down to dimming the white light as much as possible and filtering it out of the desired red of hydrogen.

A limited solution of the problem was devised in 1868. By attaching a spectroscope to a telescope, bringing the slit of the spectroscope tangent with the edge of the sun and moving the slit in a circle around the edge, prominences could be detected. Then the slit could be opened wide enough to see an entire small prominence or part of a large one. Even though this method distorts the image of a prominence, it is still used on special occasions.

In 1890 George Ellery Hale and Henri Deslandres independently invented the spectrohelioscope. This instrument utilized the red light of hydrogen to produce an image of the entire disk of the sun. Thus prominences were visible not only at the edge of the sun but also on the face of it. For 40 years the spectrohelioscope was the major source of knowledge about solar prominences.

The sun is entirely gaseous. No solid surface lies beneath its outer layers. From the outside in, the outer layers are the corona, the chromosphere, the reversing layer and the photosphere (the layer we see when we look at the sun with the naked eye or a conventional telescope). An orderly mind instinctively seeks to arrange these layers in a diagram resembling the cross section of an onion. Illustrations of this sort do not appear in textbooks, largely because they would convey the misleading impression that the layers have sharp boundaries.

The corona, the pale yellow and pearly white outer aura seen during a total eclipse of the sun, is by far the thickest of the four layers. It extends beyond the visible disk of the sun for about one third of its diameter, and sometimes much farther. It is a diaphanous thing, fainter than moonlight. The scarlet clouds of hydrogen that comprise the prominences appear to shoot up from the chromosphere, which contributes only 12,000 miles to the 864,000-mile diameter of the visible solar disk.

Although the prominences take various flamelike forms, they are not flames in the ordinary sense because the sun does not burn. The astronomer

Edison Pettit classified and named the prominences according to their behavior. There are three active types: "interactive," "coronal active" and "common eruptive." The latter are subdivided into "quasi-eruptive," "common eruptive," and "eruptive arch." Then there are the sunspot types: "cap," "common coronal sunspot," "looped coronal sunspot," "active sunspot," "surge," "ejection," "secondary" and "coronal cloud." The tornado types are "columnar" and "skeleton." Finally there are the "quiescent" and "coronal" types, which have no variants. Sometimes 20 prominences of these various types are simultaneously visible at the edge of the sun. Sometimes there is no prominence for days. Prominences erupt at velocities up to 451 miles per second; a common velocity is 100 miles per second. They often change from one type to another.

The natural fascination of watching things that move may largely explain the amateur astronomer's dream of owning an instrument that makes prominences visible. Because the prominences are so large by terrestrial standards, they appear to move lazily. But like the slow movement of the hour hand of a clock, their transformations can easily be perceived over a matter of minutes. When recorded by time-lapse photography, and then projected as a motion picture, prominences are an awesome spectacle of nature. Henry Paul of Norwich, N.Y., chemist and amateur astronomer, has written: "Spine-tingling excitement tinged with awe usually accompanies the first viewing of an eruptive prominence. These millions of tons of glowing gas often stream back in graceful arcs to the surface as if drawn by a huge magnet."

Paul's exciting experience came not out of a spectrohelioscope but from a newer instrument called the coronagraph. Where the spectrohelioscope sorts out the red light of hydrogen by means of a diffraction grating, the coronagraph does so with a remarkably selective filter that only passes light emitted at the hydrogen-α transition.

In general the coronagraph is the better instrument. One astronomer who began as an amateur telescope maker and has used both instruments says that the views with the coronagraph are enough to make spectrohelioscope observers feel that their lives have been wasted. Another calls this an understatement. The older instrument does enjoy some advantages. It is superior, for example, in revealing details of the chromosphere, unless the design of the coronagraph is carried to the extreme limit of its capability—at greatly added cost. Moreover the spectrohelioscope may be adjusted to filter light of any color.

The coronagraph was independently invented by several astronomers: first the late Bernard Lyot of France and later Yngve Ohman of Sweden and John S. Evans of the U.S. (who began as an amateur telescope maker).

However, this instrument shown here was devised by an amateur, Walter J. Semerau, in Kenmore, NY.

After using his instrument for a year, Semerau wrote: "My coronagraph performs beyond all my expectations. Building it was the most fascinating fun I have ever had."

Semerau mounted his coronagraph on the same telescope axes that support a 12½-inch astrographic camera he had made earlier. At the bottom of the coronagraph tube a pair of right-angle prisms jackknifes the light beam from the long-focus objective. Hence what appears to be two parallel tubes is in effect a very long single tube folded for the sake of convenience. In one part of the tube the light passes through a rectangular box containing the filter. Semerau built his first coronagraph in 1954. Back then he had to fabricate his own filter using a collection of polished quartz rods and plastic polarizing films. Fortunately, today's amateur can purchase extremely good H-alpha filters commercially. Search the Web for "solar filters" to find out what's available, or see the supplier list on page 255. The filters come in pairs, with an energy rejection cover that attenuates the light that enters the telescope to prevent overheating, in the H-alpha filter itself. These filters are temperature sensitive. So to keep things working perfectly, they usually come in their own temperature-controlled boxes. Experienced amateurs who can build their own temperature-controlled enclosures may be able to save some money by purchasing the filter separately.

The monochromatic light emitted by the prominences is passed upward through the filter and into the eyepiece by a reflex mirror. Light of other colors is absorbed by the filter and dissipated in the form of heat. The light can be directed into either of two cameras by moving a small lever attached to the mirror. The instrument can thus serve as either a coronascope or a coronagraph. The arrangement enables the observer to view prominences until a photograph is needed. The drawing shows a 35-millimeter camera in place. It may be replaced by a time-lapse camera or a CCD camera.

The business end of the coronagraph is the *quartz polarizing monochromator*. Though it is a filter, the term is misleading because it suggests simple glass or gelatin filters. Even the best of these would transmit a band of color much too broad. The filter must exclude all light waves which do not fall within a single hydrogen line—a band of about three to seven angstrom units in the 3,750 angstroms of the visible spectrum. To accomplish this the *quartz polarizing monochromator* sends the light through a stack of six or more *quartz plates*. At the top and bottom of the stack and between each pair of plates are sheets of Polaroid. In addition to this filter the coronagraph requires a cone of metal (the circular bottom of which

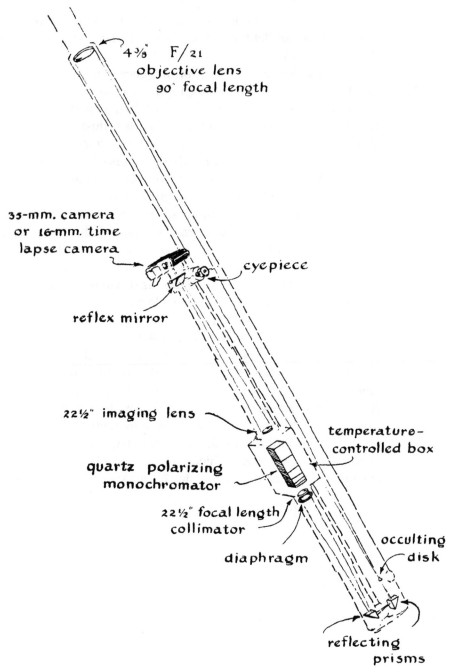

Figure 13.1 The optical system of the coronagraph

103

eclipses the sun in the instrument), a field lens, a diaphragm, a lens to send the rays parallel through the *quartz plates* and another lens to focus the emerging rays.

Since the filtered light of a prominence is narrowly monochromatic, the objective lens of the coronagraph need not be corrected for color. In fact, a single-element lens works better than an achromat. The lens must be superlatively free of bubbles, striae, dirt, dust and even microscopic scratches which would diffract light. Single-element lenses may also be used beyond the eclipsing cone; these need not be so free from defects. A reflecting telescope cannot be used for a coronagraph because irregularities in the metal reflecting surface diffract light.

Semerau writes: "Since most solar observatories are perched on high mountains in clean air far from the smoky industries, many have the belief that very little can be done at a solar observatory at low altitude among industries. It is true that the higher the observatory and cleaner the air the better, but the amateur astronomer will be amazed at what he can see and learn of our sun from such a locality.

"A year of observing experience in the heavily industrialized area of Buffalo, N.Y., at an altitude of only 600 feet, has taught me the following facts about the coronagraph. A clear blue sky, preferably after a rain, with temperature not above 72 degrees, will give finest observing and photography. From mid-July to mid-September observing must be done before 10 a.m. while the earth is still cool. After mid-morning the heat from city streets and industry makes observing impossible. Observing is generally bad during December, January and February, when the sun is low and the light must pass through a thicker layer of smoke and haze. Prominences are just barely visible on a cool day if the sky is gray-blue or hazy."

14 TWO SPECTRO-HELIOGRAPHS TO OBSERVE THE SOLAR DISK

Conducted by C. L. Stong, April 1958 and March 1974

Some years ago a young coal miner in West Virginia sent a letter to this department which began: "Those who helped make the amateur-telescope-making books possible have caused me to live two years of my life in complete contentment." The letter went on to tell how its author, Walter J. Semerau, of Kenmore, N.Y., had constructed a six-inch reflecting telescope. In the intervening years Semerau had a remarkable career both in amateur optics and in his daily work. He left coal mining to become an electrician, then an instrument-maker, then a laboratory technician and finally an engineer. Meantime, his six-inch telescope has been succeeded by a whole galaxy of bigger and better instruments, including a 12½-inch reflector complete with a coronagraph and a spectrograph. Semerau now informs us that his telescope mounting supports a new apparatus which has long been the dream of amateur telescope makers; a spectroheliograph of the Hale type. This instrument provides him with a view of the sun rarely enjoyed by laymen.

"Although the sun is a fairly stable body of gas," writes Semerau, "it is neither as amorphous nor as placid as the casual viewer might suppose. Immense clouds of ionized hydrogen, calcium and other substances thrown up from the interior account for features as distinct as the earth's oceans and land masses. Although each of these features emits light of unique color and intensity which distinguishes it from its surroundings, they are lost in the white glare of the sun as it is seen by the naked eye. To see the details clearly the observer is obliged to examine the sun in light of a single color.

"One might suppose offhand that the details could be brought into view by looking at the sun through a filter of colored glass. This stratagem would fail because colored glass, however deeply it is stained, transmits a broad band of colors, just as a radio set of poor selectivity permits several broadcasting stations to be heard at the same time. One must use a filter with an extremely narrow 'pass-band.'

"Such a device was hit upon about a century ago in India by the French astronomer Pierre Janssen. Janssen was using a spectrograph equipped with two slits to observe a total eclipse of the sun. The image of the sun's edge was focused on one slit. Rays proceeding through the slit were spread out by the prism into the familiar ribbon of spectral lines. Janssen was examining one of the lines through the second slit—the dark red line characteristic of glowing hydrogen—when he saw a tongue of flame standing out from the solar edge. Opening the slit brought more and more of the prominence into view until the width of the slit exceeded that of the red line. At this point the image became blurred. To examine slit-shaped portions of the solar disk in other colors Janssen simply shifted the viewing slit to other lines of the spectrum.

"Some 40 years later George Ellery Hale and Henri Deslandres independently devised a method of using the double-slit spectrograph to make photographs of the whole solar disk. The two slits were simply coupled mechanically so that they could be moved as a unit. When the entrance slit is swept across the sun's image, the exit slit keeps in step with the similarly moving spectral line of any selected color. Solar features emitting light of that color are focused on a photographic plate and build up a composite image that resembles the scanned image of a television picture. The device, called the spectroheliograph, was only a step away from the spectrohelioscope, which presents the composite image to the eye. To make a spectroheliograph into a spectrohelioscope one simply arranges for the slits to oscillate across the sun's disk at a rate of 20 or more sweeps per second and substitutes an eyepiece for the photographic plate.

"Not many spectroheliographs have been built by amateurs because of the difficulty of procuring the element which disperses white light into its constituent colors. This may be either a glass prism or a diffraction grating.

"Although it is possible to fit out a spectroheliograph for mechanical scanning, the arrangement is bulky, difficult to maintain and a remarkably effective generator of unwanted vibrations. For these reasons I adopted the optical-scanning system devised by Hale in 1924. The conventional rocker arm which carries the entrance and exit slits in the mechanical system is replaced by a pair of rotating glass cubes, or Anderson prisms. The image of the sun is focused on the fixed entrance-slit of the spectrograph through one prism. The image of the similarly fixed exit-slit is focused on the plateholder (or on the focal plane of the eyepiece) through the second prism. As the prisms rotate, refracted light sweeps past the slits as though the slits had been moved across the rays mechanically. The prisms are mounted on the ends of a shaft which turns on ball bearings; the unit can be assembled on the mounting of even a small telescope without introducing perceptible vibration.

"The spectroheliograph assembly consists of (1) a main housing to which the moving parts are attached and (2) a tube for the eyepiece, reflex mirror and 35-millimeter camera, CCD camera, or CD camera [*see drawing on page 108*]. The unit is relatively light, compact and simple in construction. It weighs 10½ pounds complete with eyepiece and camera, and measures 15 inches over all. An adapter makes the assembly interchangeable with the plateholder of the spectrograph, which is mounted beside the telescope. The bearings of the equatorial mounting have enough friction to offset the added weight of the unit; thus no change is required in the counterbalance when the spectroheliograph is used.

"Each prism is clamped between a flange at the end of the shaft and a metal disk held in place by through bolts. Rubber sheeting between the glass and metal protects the prisms against excessive mechanical strain. Center to center the prisms are 3.625 inches apart, the distance between the entrance and exit slits of the spectrograph. The flange supporting the outer prism is grooved for an 'O' ring belt through which the rotating assembly is coupled to a miniature direct-current motor. The facets of the prisms must be adjusted to lie in a common plane or the image will flutter when the unit is started up.

"Parallel rays entering my telescope come to a focus at a distance of 62.5 inches from the 12.5-inch objective mirror, a focal ratio of *f*/5. The focal ratio of the spectrograph is *f*/23. To feed the spectrograph with light from the telescope a set of negative achromatic lenses was introduced into the optical path between the objective and the spectrograph. This compensates for the difference between the focal ratios of the two instruments. A pair of front-silvered optical flats was mounted at the upper end of the telescope to receive rays reflected from the objective and bend them 180 degrees into the spectrograph.

"Incoming rays pass through one rotating Anderson prism, scan the entrance slit and diverge to an eight-inch spherical mirror at the opposite end of the spectrograph tube, where they are reflected as parallel rays to

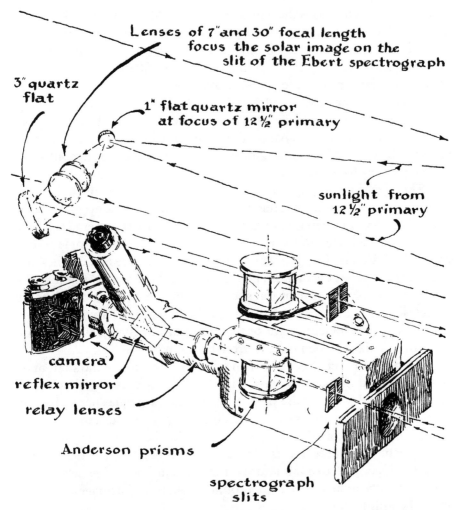

Lenses of 7" and 30" focal length
focus the solar image on the
slit of the Ebert spectrograph

3" quartz
flat

1" flat quartz mirror
at focus of 12½" primary

sunlight from
12½" primary

camera
reflex mirror
relay lenses

Anderson prisms

spectrograph
slits

Figure 14.1 Details of Semerau's spectroheliograph

the diffraction grating at the other end of the tube. Here the white light is
dispersed into its component colors and returned to the spherical mirror,
which brings the resulting spectrum to a focus in the plane of the exit slit.
The exit slit may be adjusted to match the width of any spectral line. The
most useful lines are the red 'alpha' line emitted by glowing hydrogen and
the 'H' and 'K' calcium lines in the violet region of the spectrum. Light
transmitted by the exit slit proceeds through the second Anderson prism,
the scanning action of which, together with a final lens assembly, recon-
stitutes a highly monochromatic image of the source in the focal plane of

the camera. A reflex mirror in the beam permits the image to be examined visually through the eyepiece.

"The operating procedure is relatively simple. After the instrument is assembled and aligned, the angle of the diffraction grating is adjusted to bring the desired spectral line into view in the eyepiece.

"Diffraction gratings produce a series of spectra (spectral orders), an effect analogous to a multiple rainbow. The extent to which the colors are dispersed increases with the 'higher' orders at the cost of brightness. When used in the first order, the grating of my spectrograph can spread 14.5 angstroms of the spectrum enough to fill the exit slit when its jaws are spaced one millimeter apart. In other words, dispersion in the first order is 14.5 angstroms per millimeter. The second order gives a dispersion of 7.5 angstroms per millimeter, and the succeeding orders proportionately more. Thus, were it not for the fact that the brilliance of the diffracted light diminishes with each successive higher order, one could observe an extremely narrow band of color through an exit slit of substantial width. My grating is ruled for use in the second order (it is 'blazed' for 10,000 angstroms in the first order and 5,000 in the second). Hence, to observe a band of color one angstrom wide the jaws of the exit slit must be spaced about a seventh of a millimeter apart.

"The spectral orders produced by the grating tend to overlap; that is, the red end of the first order falls on the violet end of the second, the red end of the second order overlaps the violet end of the third, and so on. The effect must be minimized or the quality of the final image will suffer. This is accomplished by inserting in the optical path a glass filter which has maximum transmission in the region of the spectrum under observation. If one is observing the alpha line of hydrogen, for example, violet light from the unwanted order will be suppressed by a red filter such as the Corning Glass Works' No. N1661. Similarly, a violet filter is used when observing the H or K lines of calcium. The filter may be inserted at any point in the system, but a filter located at or near the primary focus will heat unevenly and may break. Hence the filters are usually placed at a point between six and eight inches from the primary focus.

"With the filter in place, the entrance slit is opened to a width of about .02 inch. This admits considerably more light than is needed for observing but simplifies subsequent adjustments. The exit slit is opened so the spectral lines can be seen easily between the jaws. The desired line is selected and focused sharply by moving the entrance slit back and forth. The image of the exit slit is then focused so that the jaws appear sharp when viewed through the eyepiece. If spectroheliograms are to be made, the camera is similarly adjusted to bring the slit into sharp focus on the film. The instrument is next adjusted for maximum resolution. First, the jaws of the entrance slit are closed to the point where the spectral lines appear dark and sharply defined against a light background. Then the jaws of the exit slit are closed until they just frame the line selected. In the case of the alpha line of hydrogen the optimum width will be approximately a fifth of a millimeter. The motor is started. As the prisms reach a speed of about 16 revolutions per second, a monochromatic image of the sun, complete with the flaming detail of the solar surface, will come into view.

"This of course assumes that all adjustments have been carefully made. Each element of the instrument, from the objective to the eyepiece and camera, must be aligned with the optical axis of the system. If the telescope and spectrograph are out of line, for example, only part of the light will fall on the diffraction grating. The final image will not be as bright as it could be. In addition, the unused light will be reflected from the housing, will mix with the diffracted rays and reduce the contrast of the image. Similarly the system should be adjusted so that white light from the entrance slit approximately fills the diffraction grating. If the grating is not fully illuminated, its efficiency suffers. Conversely, rays which extend beyond the edge of the grating are lost to the final image and impair its contrast.

"The final adjustment consists in gradually narrowing the exit slit. This brings progressively finer details into view; prominences, mottling near the region of sunspots, dark filaments, flocculi and so on. It also reduces the brilliance of the image and sets a limit to visual observation. At this point the camera comes into use. The average exposure time is from two to four seconds; the instrument is guided during a time exposure as it is in conventional astronomical photography. The camera is also used in the violet region of the spectrum beyond the range of the eye. In this region lie the H and K lines of calcium.

"Although the instrument has many desirable features, I should also mention one disadvantage. The Ebert spectrograph, as I constructed it, introduces some distortion; the image of the sun's disk is somewhat elliptical. This is explained by the fact that the slits must be located somewhat off the axis of the spherical mirror. The distortion is partially compensated by tilting the camera. Curved slits would provide a better correction, but I have no way of making them. The distortion does not impair resolution but it introduces some complication in locating details accurately on the image. Advantages of the design include simplicity, lightness, relatively low cost and a cylindrical form that is easy to assemble on an equatorial mounting. In addition, desired portions of the spectrum can be brought into view at the twist of a dial.

The spectroheliograph makes a great complement to a coronagraph [*see page 103*]. The two can be used simultaneously. The coronagraph shows prominences at the edge of the sun in great detail, but gives no hint of the solar disturbances responsible for them because the central disk is masked by a diaphragm. In contrast, the spectroheliograph reveals faculae, flocculi, filaments, spots and even prominences of exceptional brilliance.

"I had the good fortune to observe and photograph an interesting pair of solar events on October 20, 1957. No outstanding disturbance was evident when I began to observe at 14:15 Greenwich mean time, but within 15 minutes a scarlet flocculus appeared near the southwest edge of the sun. The intensity of the flocculus remained constant during the following two hours, but at 16:51 a small flare brightened near the east edge. At about this time the cloud first observed also started to brighten; thereafter both regions grew in size and brightness. By 17:15 the east flare had diminished to normal brightness and 45 minutes later the one near the southwest edge similarly faded. The visual image was sharp and crisp. Poor seeing caused some blurring of the photographs, but conditions improved somewhat at 16:51."

Gene F. Frazier of Washington, D.C. views the sun routinely with a different homemade instrument. In certain respects Frazier's apparatus resembles the spectrohelioscope just described. His instrument has an additional diffraction grating but requires no solar telescope or motor-driven optical parts. He describes the principles, construction and operation of the apparatus as follows.

"Essentially the instrument employs an external diffraction grating to disperse and reflect sunlight to a concave mirror. The mirror projects the rays through an adjustable plate of flat glass to a focus in the plane of the entrance slit of a conventional spectroscope.

"A photograph that could be made by putting a photosensitive plate in the position occupied by the slit of the spectroscope would not show the dark absorption lines that normally characterize the solar spectrum. In my system the image of the sun functions as the slit. Hence a photograph is composed not of the series of narrow absorption lines but of overlapping images of the solar disk separated by distances corresponding to the wavelength of the absorbed light.

"The adjustable plate of flat glass that admits incoming light to the slit acts as a vernier for displacing the rays laterally with respect to the slit. Rays that enter the plate at an angle to its perpendicular are refracted and emerge at the identical angle. The amount of deviation is approximately proportional to the angle between the plate and the entering beam. By rotating the plate the observer can shift the spectrum any small distance with respect to the slit. The plate functions as a precision tuner that enables the experimenter to admit any narrow portion of the spectrum to the slit.

"The selected rays, which may span a range of color only 10 angstroms wide, emerge from the slit as a diverging beam. The diverging rays fall on a concave mirror from which they are reflected as a bundle of parallel rays to the internal diffraction grating of the spectroscope [*see illustration below*]. The internal grating disperses the colors still more. The angle at which the internal grating is set can be adjusted to reflect rays of essentially monochromatic light to the second concave mirror of the spectroscope. The second mirror reflects the rays to focus in the plane of the eyepiece.

"The details of the filtering action can be demonstrated by replacing the external grating with a flat mirror and letting sunlight fall on the mirror. After adjustment an instrument so modified would display at the eyepiece the normal solar spectrum crossed by dark absorption lines. Assume that the geometry of the diffraction grating of the spectroscope is such that each angstrom of wavelength of the solar spectrum is dispersed through a distance of one millimeter in the focal plane of the eyepiece. This was essentially the case with Janssen's spectroscope. The chromosphere of his solar image was less than one millimeter wide. Therefore he could partly isolate

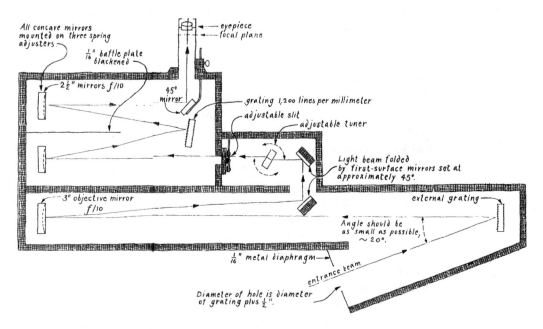

Figure 14.2 Plan view of the optical train of the filter

the emission of the prominences from background light by carefully focusing this narrow feature of the image on the slit of his spectroscope.

"Now assume that the flat mirror is replaced by the external diffraction grating of my instrument and that the angle of the grating is carefully adjusted to reflect a narrow band of light on the slit that spans 10 angstroms (from, say, 6,558 to 6,568 angstroms). The spectrum is noncoherent. For this reason the light that reaches the slit consists of many monochromatic images of the sun's disk that overlap on each side of the hydrogen line at 6,562.8 angstroms. If the dispersion of the gratings is assumed to be one angstrom per millimeter, the centers of each of the images of the sun's disk would be separated by one millimeter. At any setting 10 solar disks would overlap.

"This means that a band of color only 10 angstroms wide can enter the slit and that the scattering of light is significantly reduced. When the instrument is adjusted for observing prominences at 6,562.8 angstroms, unwanted light is reduced by more than 95 percent! Indeed, on a clear day it is not unusual for the field to appear completely dark five angstroms from the image. The full solar image appears in the field of view, which helps the observer to keep the edge of the image centered on the 6,562.8-angstrom line. With the aid of the tuner I have easily observed prominences continuously for intervals of more than two minutes.

"The construction requires no special skills, but the quality of the gratings is crucial. They must be mounted with care. The gratings should be ruled with at least 1,200 lines per millimeter for adequate resolution and high dispersion. The ruled area of the gratings in my instrument measures two inches square. The lines are blazed for 6,600 angstroms in the third order, which is to say that the surface of the rulings is cut at an angle that reflects light of maximum intensity at the 6,600-angstrom wavelength in the same direction as the grating disperses these rays in the third order.

"The gratings can be mounted in simple structures bent in the form of a V from sheet steel or brass. The gratings can be attached lightly to these mountings with machine screws and insulated from the metal with felt lining. The metal V's are supported by soldering the rear side to a quarter-inch copper rod that fits a radio dial of the vernier type [*see illustration on page 114*]. The copper rod must be bent to an angle such that the projected axis of the vernier dial bisects the plane of the grating. When the rod is so mounted, the angle of the gratings with respect to the impinging rays can be altered without displacing the spectral orders at the eyepiece.

"I made adjustable cells of plywood for supporting the mirrors. The cells are supported at three equidistant points by machine screws fitted with compression springs and wing nuts. The mirrors can be lightly fastened to the ply-

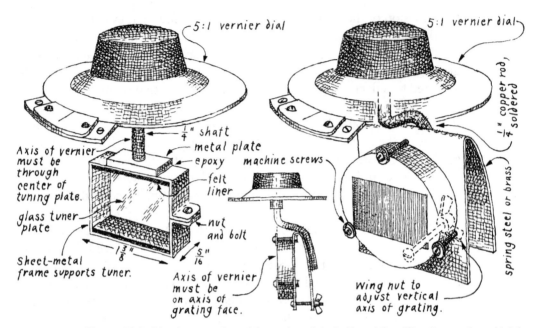

Figure 14.3 Vernier mounting of the tuning plate (left) and the diffraction gratings (right)

wood with wood screws insulated by rubber tubing and fiber washers. Incidentally, adjustable cells of cast aluminum are available commercially at a reasonable price for mirrors of three-inch diameter or more. The cells also accept two-inch mirrors mounted in three-inch washers of plywood.

"The diameter of the mirrors is not critical, but it must be at least as large as the ruled area of the gratings to prevent vignetting (obscuration at the edges of the image) and the scattering of stray light into the image. In addition the focal length of the objective mirror should be an integral multiple of the focal length of the spectroscope mirrors, which, in turn, should be equal to within a tolerance of about 2 percent. The quality of the final image can be optimized by mounting the two mirrors of the spectroscope as close together as possible in order to minimize off-axis aberrations.

"The tuning plate can be made of any optically flat glass about 50 millimeters wide and six to 10 millimeters thick. The piece can be circular or rectangular. Plates of this size that were originally intended for use as optical windows are now available inexpensively from dealers in surplus optical supplies. The plate can be mounted by a frame of sheet metal and adjusted by a supporting shaft and a vernier dial.

"The construction of an adjustable slit of adequate quality has through the years remained the most difficult problem that confronts amateurs who make spectroscopes. The best slits by far are the ones that can be bought from distributors of optical supplies, but they are currently expensive. The slit must remain rigidly centered when its width is increased from zero to two or three millimeters, which means that both jaws must move equal distances in opposite directions when the device is adjusted.

"A mechanical linkage in the form of a parallelogram can satisfy this condition. The system of links can be assembled with snugly fitting machine screws. Excess play in the system can be eliminated by inserting a pair of helical springs to maintain a few grams of tension between the side links that support the jaws of the slit. The jaws can be made of single-edge safety-razor blades, which can be fastened to the supporting links with epoxy cement.

"I prefer jaws made of sections cut from a hacksaw blade. I first grind off the saw teeth with a carborundum wheel. The opposite edges are polished to remove surface irregularities. The procedure is not difficult. I grind two four-inch slabs of plate glass together with a thin slurry of No. 120 carborundum grit in water for a period of six minutes, making elliptical strokes about an inch long and turning the 'sandwich' over every minute. The edges of the blade are ground against the frosted side of either of the glass pieces for two minutes, again with elliptical strokes. I examine the edges for pits and hills and, if necessary, continue grinding until they are straight and smooth.

"One of the completed jaws is soldered or cemented with epoxy to its supporting linkage. After the jaw has been fastened the linkage is moved to the position where the separation of the side links is at a minimum. The ground edge of the companion jaw is placed in full contact with the ground edge of the jaw previously installed and similarly attached to its supporting link.

"All optical elements of the instrument must be installed in a light-proof housing. I improvised one of plywood. The spectroscope was made as a separate unit that could be bolted to the housing that supports the external grating and the objective mirror. This arrangement enables me to employ the instrument as a conventional laboratory spectroscope.

"The housings can take the form of simple boxes with removable lids to which the vernier dials are fixed. I minimized the overall dimensions of the apparatus by inserting a pair of plane front-surface mirrors between the objective mirror and the tuner to fold the incoming rays. The plane mirrors are set at an angle of approximately 90 degrees with respect to each other. They must be supported by mountings that can be adjusted a few degrees to align the optical path. Another plane mirror similarly mounted reflects converging rays from the spectroscope to the focal plane of the eyepiece. The interior of the spectroscope housing, including particularly the barrier that separates the concave mirrors, should be painted flat black to minimize the reflection of scattered light.

"All optical parts should be tested before they are installed. Testing the focal length of the mirrors is particularly important. A simple measurement of the focal length can be made by standing the mirror on edge, directing a flashlight toward the metallized surface and catching the reflected image of the lamp filament on a sheet of white paper placed next to the flashlight. Vary the distance of the flashlight and the paper from the mirror until the sharpest possible image of the lamp filament appears on the paper. The focal length of the mirror is equal to exactly half the distance between the image of the filament and the surface of the mirror. The focal length can be measured with greater accuracy by setting up the knife-edge test used for checking telescope mirrors.

"The cost of this instrument will increase exponentially with the diameter of the optical parts. The two-inch gratings and mirrors have enabled me to make most of the observations I had in mind when I undertook the construction and also to do a variety of laboratory experiments.

"The instrument has a maximum dispersion of about two angstroms per millimeter, which is equivalent to displaying the rainbow as an image more than six feet wide at the focal plane of the eyepiece. The zone in which the prominences can be viewed at the edge of the sun is less than one millimeter wide, but even so it is sufficient to enable the observer to isolate the 6,562.8-angstrom spectral line of hydrogen.

"When the apparatus has been attached to an equatorial mounting that includes a clock drive to keep the ruled surface of the grating pointed approximately toward the sun, set the tuning plate at a right angle to the optical path and start the clock drive. Cover the objective mirror with a disk of white paper and adjust the external grating to the angle at which reddish light from the sun falls on the paper. Transfer the paper to the position of the slit of the spectroscope and adjust the flat mirrors to angles such that the reddish image falls on the paper at the position of the slit.

"Remove the paper. Adjust the first mirror of the spectroscope (the collimating mirror) to the angle at which the now parallel rays of the reddish light flood the diffraction grating of the spectroscope. If the light is difficult to see, cover the grating with a small sheet of white paper. Adjust the second mirror of the spectroscope to project converging rays to the flat mirror adjacent to the eyepiece. A sheet of paper inserted in the focal plane of the eyepiece should now display an image of the sun.

"While viewing through the eyepiece adjust the angle of both gratings to center the dark 6,562.8-angstrom spectral line on the slit of the spectroscope. The line is the darkest and broadest one in this region of the spectrum. The observer should now see in the eyepiece the complete scarlet image of the sun.

"To observe the limb, rotate the tuner so that the image appears to shift along the absorption spectrum to the point at which the edge of the sun is centered on the line at 6,562.8 angstroms. That is the adjustment at which Janssen made his initial observation. If a prominence happens to be located at this point, the observer will see the absorption line fade and be replaced by a bright area.

"Only a small area of the sun's edge is being observed. For this reason the edge will appear to curve away from the straight slit. To see prominences along a substantial portion of the edge the straight slit must be replaced by one that matches the curvature of the image.

"Curved slits of two kinds are relatively easy to make. Of the two I prefer one that requires the use of an engine lathe to cut a disk of metal equal in diameter to the diameter of the solar image at the focal plane of the objective mirror. The dimension is very nearly equal to the focal length of the objective divided by 114.

"In the case of an objective mirror with a focal length of 30 inches the radius of the curved slit is approximately .131 inch. A hole .05 inch larger in radius is drilled in a metal sheet. The device is assembled by centering the disk in the hole and tacking it in place with a dab of solder or epoxy cement to leave a clear arc .05 inch wide and extending about 180 degrees.

"Another technique for making the curved slit is easier. Drill a hole slightly larger in diameter than the solar image and place it over a mirror that has been aluminized or silvered. Dip the sharpened tip of a wood toothpick in dilute nitric acid. Shake excess acid from the wood. Insert the sharpened tip through the hole in the metal and, with the metal as a guide, trace a semicircle on the coated glass. If the operation is performed with care, the acid will etch a usable slit in the metallic film. To observe solar prominences substitute the curved slit for the straight one.

"In a sense the double-grating filter is analogous to the tuning system of a radio set. It enables the investigator to select for observation a narrow band of light waves much as the dial of a radio set tunes in a narrow interval of the radio spectrum. The filter can serve in a variety of experiments other than the observation of solar prominences. For example, my interests include the use of polarimetry for investigating the characteristics of minerals. At various orientations the surface of a rock can be seen in different colors which depend on the crystal structure of the specimen. By examining the rock with the tuned filter, with both gratings adjusted to an appropriate angle, distinct colors appear in various areas of the surface that characterize the specimen.

"Mineralogists also routinely dissolve bits of unknown rock in a bead of incandescent borax to identify its constituents by the characteristic colors of the resulting flame. Each chemical element radiates a unique set of wavelengths. With the double-grating filter the experimenter can observe and even photograph the distribution of elements in the glowing gases.

"The rulings of a grating are cut at an angle that optimizes the efficiency of the device as a reflector of light of specified wavelength at a specified angle with respect to the plane of the rulings. As I have mentioned, the angle at which the rulings are cut is called the blaze. The angles at which gratings reflect bundles of rays dispersed in the form of spectra are known as the spectral orders.

"As I have mentioned, the gratings of my instrument are blazed to reflect most of the incident light at 6,600 angstroms in the third order. In general dispersion increases with the spectral order at the cost of brightness. My gratings were selected primarily for viewing solar prominences. Hence they were blazed for maximum brightness of the deep red in the spectral order that resulted in a dispersion of two angstroms per millimeter at the focal plane of the eyepiece. People who design the instrument for experiments of other kinds such as flame spectroscopy would doubtless select gratings blazed for other colors in other spectral orders.

"Two people have helped me with this study. Timothy O'Hover of the chemistry department of the University of Maryland provided a laboratory and Gary A. Frazier supplied inspiration and electronic testing equipment for the original studies. I am deeply grateful to them."

PART 3

THE EARTH,
MOON,
AND
SATELLITES

15 A PENDULUM THAT DETECTS THE EARTH'S ROTATION

Conducted by C. L. Stong, June 1958

One day in the middle of the 19th century the French physicist Léon Foucault inserted a slender rod in the chuck of a lathe and plucked the free end of the rod so that it vibrated like a reed. When he started the lathe, he observed that the turning rod continued to vibrate in the same plane. Later he tried much the same experiment with a vertical drill-press. From the chuck of the drill-press he suspended a short pendulum consisting of a length of piano wire and a spherical weight. He set the weight to swinging and started the drill-press. The pendulum also continued to vibrate in the same plane. For a time nothing much came of these observations, but in them was the seed of an experiment destined to settle a classic scientific controversy.

In the year 1543 Nicolaus Copernicus had sent a copy of his new book, *On the Revolutions of the Celestial Orbs,* to Pope Paul III with a note containing a historic understatement: "I can easily conceive, most Holy Father," he wrote, "that as soon as people learn that in this book I ascribe certain motions to the earth, they will cry out at once that I and my theory should be rejected." Cry out they did, and some, including a few scientists, were still crying out in 1850, when Foucault was invited to stage a science exhibit as part of the Paris Exposition scheduled for the following year. Being not only a gifted physicist but also something of a showman, he selected as the site of his exhibit the church of Sainte Geneviève, also known as the Panthéon.

From the dome of the Panthéon he hung a pendulum consisting of 200 feet of piano wire and a 62-pound cannon ball. On the floor, immediately below the cannon ball, he sprinkled a layer of fine sand. A stylus fixed to the bottom of the ball made a trace in the sand, thus recording the movement of the pendulum. Great care was taken during construction to exclude all forces except those acting vertically to support the system. Tests were even made to assure symmetry in the metallurgical structure of the wire. Finally the ball was pulled to one side and tied in place with a stout thread. When the system was still, the restraining thread was burned.

The pendulum made a true sweep, leaving a straight trace in the sand. In a few minutes the thin line had expanded into a pattern resembling the outline of a two-bladed propeller. The pattern grew in a clockwise direction, and at the end of an hour the line had turned 11 degrees and 18 minutes. This could be explained only on the basis that the earth had turned beneath the pendulum. Copernicus was vindicated.

It is easy to visualize what had happened if one imagines a pendulum erected at the North Pole. The pendulum is hung, perhaps from a beam supported by two columns, in line with the earth's axis. The supporting structure corresponds to Foucault's lathe or drill-press—or the dome of the Panthéon. So long as the pendulum is at rest the whole affair simply turns with the earth, making one complete revolution every 23 hours and 56 minutes (the sidereal day). The pendulum is now drawn out of plumb and carefully released. The direction of the swing persists, and the earth turns beneath the pendulum. To an observer on the earth the plane of vibration appears to rotate in a clockwise direction, because at the North Pole the earth turns counter-clockwise. The same effect would be observed at the South Pole, except that there the rotations would be reversed. If the swing of the pendulum appears to turn clockwise at the North Pole and counter-clockwise at the South Pole, what happens at the Equator? Substantially no deviation is observed. Here the entire system—the earth, the supporting structure and the pendulum—is transported almost linearly from west to east [*see illustration on page 123*].

At the poles the rate at which the swing of the pendulum appears to rotate is 15 degrees per sidereal hour; at the Equator it is of course zero degrees per sidereal hour. The rate varies with latitude; the higher the latitude, the higher the rate. This neglects certain fine deviations caused by the curvature of the earth's orbit around the sun and by other perturbations. But it holds for the gross, easily observed, motion.

With the help of a few simple geometrical concepts it is easy to see why a pendulum appears to rotate more slowly as it approaches the Equa-

plane of oscillation of a pendulum at the earth's pole appears to rotate once a day

plane of oscillation of a pendulum at the Equator appears to be stationary

Figure 15.1 Pendulums at the North Pole and Equator

tor. Imagine an arc along an intermediate parallel of latitude through which the earth has turned during a short interval, say an hour or so. The angle subtended by the arc at the earth's axis increases at the rate of 15 degrees per hour—one full turn of 360 degrees per sidereal day. Now assume a pair of tangents to the earth's surface subtended by the arc which meet on a projection of the axis at a point in space above the pole. The angle between the tangents increases in size at the same rate with which the pendulum's plane of vibration appears to rotate. The reason it does so becomes clear if one assumes that the pendulum continues to vibrate in the plane of the first tangent as it is transported to the position of the second. At the pole the two angles are equal; both increase at the rate of 15 degrees per hour. At the Equator the angle subtended by the arc at the earth's axis continues to increase at the rate of 15 degrees per hour. But the pair of tangents subtended by the arc at the Equator meet the projected axis at infinity. The angle vanishes, and its rate of increase does the same. Therefore the pendulum's rate of rotation is zero. Foucault demonstrated that the apparent rotation of the pendulum varies with the trigonometric sine of the latitude at which it is installed. Its rate at points between the poles and the Equator is equal to 15 degrees per hour multiplied by the sine of the latitude.

Amateurs who set up pendulums at New Orleans will observe an apparent rotation of about 7 degrees and 30 minutes per hour. They must wait two days for a full revolution. Those in Manila must wait four days; those on Howland Island in the South Pacific, about 40 days!

About the best way to gain an understanding of the Foucault pendulum is to make one. Like many such enterprises, this seems simple until it is tried. Many amateurs who have felt the urge to set up the apparatus have abandoned the idea because they had no Panthéon in which to hang a 200-foot pendulum and no cannon ball for a bob. This is no problem. Pendulums 10 or 15 feet long can be made to work handsomely with bobs weighing as little as five pounds. The most vexing problems encountered in making a pendulum have to do not with its size but with starting the bob in a true swing, maintaining the trueness of the swing, and supplying energy to the bob.

Foucault's method of starting the bob is still the most elegant. Many starting devices have been tried: mechanical releases, magnetic releases, mechanisms which accelerate the bob from dead rest, and so on. It is generally agreed that burning a thread is the simplest and best of the lot.

For years the problem of making the pendulum swing true resisted some of the world's best instrument makers. It seemed clear that any method of suspension must have radial symmetry. To assure this Foucault and subsequent experimenters took great pains in procuring wire of uniform characteristics and in designing the fixture to which the wire was attached. Roger Hayward, the illustrator of this department, tells me that the wire for the Foucault pendulum in the Griffith Observatory in Los Angeles was specially drawn and tied to a long two-by-four beam for shipment from an eastern mill to the West Coast. The designers were afraid that coiling the wire would destroy its symmetry. The pivot to which the wire was attached at first consisted of a set of gimbals with two sets of knife-edges at right angles to each other. Despite these precautions the completed pendulum insisted on performing figure eights and ellipses. Hayward, who had designed other exhibits for the Observatory, suggested that the wire simply be held in rigid chuck. This invited a break at the junction of the wire and the chuck, which could cause the wire to lash into a crowd of spectators. To minimize this hazard a crossbar was clamped to the wire just above the ring-shaped driving magnet. Thus if the wire had broken, the crossbar would have been caught by the magnet ring. Clamping the wire in a chuck cured the difficulty. The wire has now been flexing for more than 60 years without any apparent ill effect.

At about the time the Griffith pendulum was installed a French physicist named M. F. Charron devised a method for maintaining the true swing even when the forces acting on the pendulum are measurably asymmetrical. Charron set out with the objective of designing a vise which would grip the wire rigidly and would also provide a long radius through which the wire could flex. He used a ferrule which at its upper end fitted snugly around the wire and at its lower end flared away from the wire [*see illustration on page 125*].

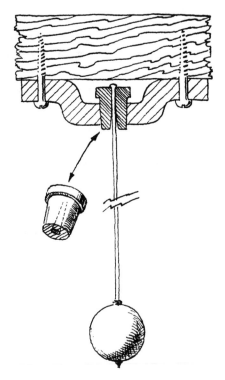

Figure 15.2 Charron pivot for a
Foucault pendulum

The diameter of the hole at the lower end precisely accommodated the swing of the pendulum. It was observed that when the pendulum swung true, the wire simply made contact with the inner surface of the ferrule at the end of each beat. But when the pendulum performed ellipses or other configurations, the wire rubbed against the inner surface of the ferrule and the energy responsible for the lateral component of motion was dissipated through friction. Stephen Stoot, a Canadian amateur, has suggested a modification for small pendulums which accomplishes the same result. He fixes a carefully centered washer around the wire just far enough below the point of suspension so that the wire touches the washer at the end of each swing [*page 126*].

The third basic problem is how to supply the pendulum with a periodic push in the precise direction in which it needs to go. No completely satisfactory mechanical solution has been devised. A variety of electrical drives are in use, however. Most of these feature a ring-shaped electromagnet which acts on an armature carried by the wire near the point of suspension. Power is applied during the portion of the swing in which the wire is approaching the magnet, and is interrupted when the wire moves away.

The arrangement in use at the Griffith Observatory is typical. Current for the ring magnet is supplied through a relay. The action of the relay, in turn, is controlled by a photoelectric cell. A beam of light, folded by mirrors so that the beam crosses itself at a right angle near the wire, actuates the photoelectric cell [*see illustration on page 127*]. When the suspension passes through the center of the ring magnet, a vane on the wire breaks the beam. The relay then operates, and applies current to the magnet. After an appropriate interval, during which the wire approaches the magnet, a time-delay relay breaks the circuit automatically.

In 1953 R. Stuart Mackay of the University of California described in

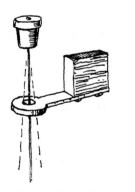

Figure 15.3 An amateur's
version of a Charron pivot

the *American Journal of Physics* a novel method of driving Foucault pendulums. His apparatus consists of a simple coil bridged by a condenser into which power is fed continuously, and over which the bob swings freely. The method takes advantage of phase-shift effects which are essential to the operation of ordinary doorbells and buzzers. It requires no contacts, light beams or other arrangements to interrupt the current.

"If one energizes a coil with alternating current," writes Mackay, "and sets an iron pendulum bob swinging immediately above it, the current through the coil will increase or decrease depending upon whether the bob is moving toward or away from the coil. The magnetic property of the iron influences the magnetic field set up by the coil and therefore the coil's inductance. The inductance, in turn, influences the flow of current. The fluctuations in the amplitude of the alternating magnetic field do not occur instantly. The effect is such that the current, and consequently the attractive force of the coil's magnetic field, is greater when the bob approaches the center of the coil than when it recedes. If the circuit is essentially inductive, enough energy will be transferred to the bob to maintain the pendulum's swing.

"The effect can be made stronger by placing a capacitor in series with the coil so that the combination resonates slightly below the 60-cycle frequency of the power line. In effect, the system is then working on one side of the resonance curve, where a given change in inductance causes a pronounced change in current.

"Some Foucault pendulums, particularly those set up as public displays, feature bobs of bronze or other nonmagnetic materials. These may also be driven by the coil. One takes advantage of the fact that such bobs can act as the secondary winding of a transformer. The magnetic field of the stationary coil and that set up by current induced in the bob are in opposition. Hence the coil and the bob are mutually repelled. As the bob approaches the coil the flow of current increases in both. But because of electrical lag in the circuits the current, which starts to rise with the approach of the bob, does not reach its peak until the bob has traveled somewhat beyond the center of the coil. Accordingly the bob is subjected to a greater net repulsive force when it is moving away from the coil than when it is approaching the coil. Energy is thus made available to drive the pendulum.

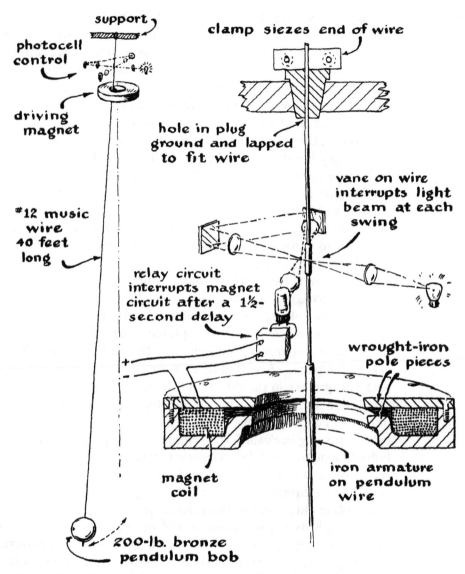

Figure 15.4 Conventional electric drive for a Foucault pendulum

"The forces acting between the coil and a nonmagnetic bob are smaller than those between the coil and a magnetic bob. Moreover, energy dissipated by currents induced in the bob causes greater damping in the system. The effect can be shown easily by substituting a short-circuited coil for the bob. The change of current in both coils, induced by changes in inductance, can be enhanced by resonating each circuit near the power-line frequency, as in the case of magnetic bobs. When the capacitor of the bob coil is made slightly too large for resonance, the circuit becomes a trifle inductive. The force between the coils is then repulsive; sustained oscillations result. If the capacitor is too small, an attractive force is produced which will not sustain the motion. It is possible, of course, to attach a driving coil tuned for repulsion to the bottom of a nonmagnetic bob.

"The first coil I tried was wound with No. 16 wire. It was eight inches in diameter, two inches thick and had a two-inch hole in the center. It was used simply because it chanced to be on hand. The coil resonated at 60 cycles when it was placed in series with a 25-microfarad paper capacitor. On the application of 10 volts it drove a three-inch iron bob on the first try. Considerably more power was delivered when the magnetic field was altered by laying the coil on a six-inch circle of $\frac{1}{16}$-inch sheet iron. Should one wish to enhance the effect further, the coil may be inserted into a cylindrical core of iron.

"Which of the two drive systems is preferable, magnetic or nonmagnetic? A performance analysis indicates little choice either way, though the magnetic bob is probably simpler to make. It is true that over a number of cycles any small perturbing force can produce a marked effect on the swing of a Foucault pendulum, usually resulting in a slightly elliptical orbit. The 'plane' of oscillation of a pendulum swinging in an elliptical path will precess in the direction of tracing the ellipse at a rate roughly proportional to its area. Thus, if one does not take care, a Foucault pendulum can appear to turn at the wrong rate, or even to indicate that the earth is turning backward. It is tempting to suppose that the nonmagnetic bob will tend to avoid the 'magnetic potential hill' that it 'sees,' that is, the repulsive force will tend to deflect the bob to one side if the swing does not pass directly through the center of the magnetic force. In contrast, a magnetic bob might appear to favor the center of the 'potential valley' it 'sees,' even though its normal swing would not necessarily bisect the field of force. Thus in either case one might expect some induced perturbation. In practice, neither has much effect on crosswise amplitude if a fairly heavy bob is used. Interestingly enough, since the motions in the two directions responsible for the ellipse are 90 degrees out of phase, and since changes in the magnetic field are controlled by the major motion, there is a slight tendency to damp out the minor motion.

"The matter of perturbing effects warrants further discussion, particularly from the viewpoint of comparing these systems with the conventional ring-magnet drive. It might seem that driving the pendulum from the top by means of a ring magnet would result in minimum sensitivity to asymmetries, but this is not strictly true. A pendulum is less sensitive to asymmetries at its top. But a greater driving force is also required there. It is the percentage of asymmetry which interests us.

"An air-core coil is, of course, simpler to construct, install and maintain than the ring-magnet drive. It must be said, however, that this method of drive does suffer somewhat in comparison with the ring-magnet system in that the whole field is not turned on and off in the course of providing useful drive. The steady useless component of the field is not necessarily symmetrical. Consequently asymmetry may be added to the system continuously. This means that in this system the magnet must be made more perfect than in a system wherein the field changes the required amount by going fully off. When desired, the magnetic field of the air coil can be trimmed for symmetry with small tabs of magnetic iron.

"For the purpose of most demonstrations, however, extreme precision is unnecessary. With reasonable care in alignment the angular velocity of the pendulum's apparent deviation should fall within 15 per cent of the anticipated value."

16 OBSERVING CHANGES ON THE MOON

Conducted by Albert G. Ingalls, June 1953

One test of the quality of a telescope is to photograph the moon with it and note how much the photograph can be enlarged without loss of sharpness. The amateur telescope maker Henry Paul has made a photograph of the moon at the focus of his 10-inch, *f*/9 reflector, where the image was eight tenths of an inch in diameter, which he was able to enlarge 20 times, giving an image of the moon 16 inches in diameter, before loss of sharpness became apparent. Usually a picture begins to be fuzzy with anything over a fivefold enlargement. Thus Paul beat par by four times.

It is easy to see the shape of a lunar crater as long as its parts cast shadows, but when the sun is high in the sky and they are illuminated on both sides, no shadows are cast. Even as large a crater as Eratosthenes can then so nearly vanish that an observer who has watched it and drawn it again and again at the telescope eyepiece may have difficulty in finding it.

The commonest method of becoming familiar with the moon's topography is to keep observing along the dark edge, because there the shadows easily interpret the relief. By doing this throughout the month, night after night as the day-night borderland creeps to the east then retreats to the west, the whole visible face of the moon is surveyed. Here many telescope users stop, and thus miss seeing changes on the moon that were long ago minutely described by the American selenographer W. H. Pickering and others. Let us ferret out some of these changes.

Pickering found it easier to observe Eratosthenes with a small telescope than Mars with a large one; the apparent diameter of Mars at its nearest is but half that of Eratosthenes, and Eratosthenes contains far more detail than Mars. To show that the major changes on Eratosthenes can be observed with a six-inch telescope anywhere in the U.S., he made drawings of it with only a three-inch telescope and a magnification of 90 from his home on the high plateau of the tropical isle of Jamaica, where the seeing is good. He systematized the study of the monthly changes by charting the eight fields in Eratosthenes as shown in the drawing on page 132. The following is his summary of the more marked changes that may be expected to be seen each month. The times given are not for a single day but simply indicate the angle at which the sun is shining on the moon at the time of observation. At 6:30 a.m. the summit of the central peak becomes visible.

At 6:40 a "canal" (streak of brightness) joining the SC and SE fields may appear.

At 7:50 "fog" (region of decreasing sharpness) begins to form within the crater, increasing till 10 a.m.

At 7:55 canals appear in the SE field.

At 8:00 the eastern side of the NW field darkens.

At 8:20 the northern part of the bird-shaped NE field begins to fade and the southern part to extend.

At 8:40 two dark spots in the E field unite.

At 9:20 the central field begins to narrow at the northern end [*see drawings on page 132, corresponding respectively with SC and C fields at 8:30, 9:30 and 10:30 a.m.*]. The northern field crosses the crater rim from the outside and begins extending inside the crater.

At 9:40 the SE field crosses the rim of the crater.

At 10:00 the SC field darkens and a canal sometimes begins joining the SC and SE fields, while the southern part of the NE field begins to fade.

At 10:40 the EC field darkens; at 1:30 it is conspicuous, and at 5:00 it is lost in the shadows.

At 10:50 the E field joins the NE, which soon begins to shrink.

At 11:00 the two arms of the SE field begin to curve inward.

At 1:00 the SC field becomes notched at the south (upper) end and at 3:20 it fades out.

At 2:00 the NE field begins to fade.

These are only the major changes. There are so many minor ones that the selenographer often has trouble keeping up with them all.

The fogs mentioned by Pickering occur in craterlets usually less than one mile in diameter. These emit a brilliant white circular glow after being warmed by the sun. They remain conspicuous until sunset. Other selenographers have reported fogs in craterlets within Eratosthenes. The British selenographer Patrick A. Moore calls them mists. He says their existence "cannot be questioned, as the evidence is overwhelming; various craters have at various times been seen mist-filled."

Pickering argued that because the markings on the moon (and on Mars as well) are neither shade nor shadow they must be either changeable surface discolorations or something growing or something moving over the surface—"mineral, vegetable or animal." He found that the "vegetation" in general is gray, like sagebrush; in places near the moon's equator it is purplish black, like lichens. Eratosthenes is one of a number of oases in the desert waste of the moon's surface. The "vegetation" is often associated with minute craterlets within large craters. Its growth and decline must be very rapid.

After Pickering published his findings in *Popular Astronomy* (November, 1919; August–September, 1921; May, 1922; February, 1924; May, 1924; August–September, 1924) critics dismissed the alleged changes as due merely to the shifting of shadows and the changing angle of illumination.

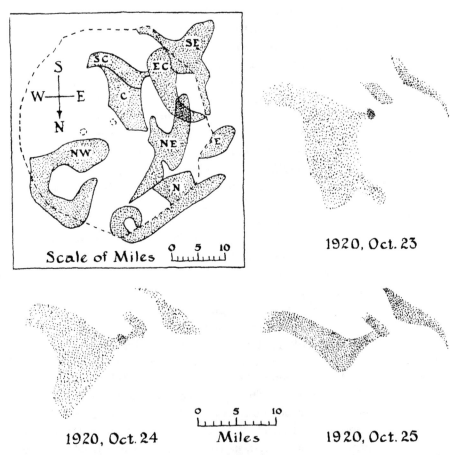

Figure 16.1 Changing areas on Eratosthenes and examples of change

Pickering replied: "If so, why do they always appear at full moon? How explain dark markings that advance *toward* the setting sun?" He denied that he had said that terrestrial vegetation would be possible on the moon. The lunar "vegetation" must be very different from that of the earth and we know nothing about what it is like. The same would be true, he said, if the dark areas were swarms of ant-sized insects migrating.

After Pickering had published his articles, the Italian selenographer Mentori Maggini spent several days discussing the moon and Mars with him and later made drawings of Eratosthenes with a 13-inch refractor in Sicily. He verified nearly all the details of Pickering's drawings. He, too, saw the canal-like streaks and noted they were continuously visible, unlike those of Mars, which peep out for no longer than one fiftieth of a second or at most a second. He said they resembled the lines of Mars and asked, "Does this similarity lead us to believe that we have to do with the same phenomenon, an optical one?"

Maggini's optical interpretation of the moon's changes was this: "The glimpsed details, or invisible ones, are gathered together and integrated into linear sensations or into spots. From the point of view of the optical theory these lines and spots are the means by which the eye of the

Spotting changing features on the surface of the moon didn't seem far fetched when these two articles first appeared back in 1953. But that was some 16 years before Apollo 11 visited the Sea of Tranquillity. We've learned a lot about the moon since then. Today we know from the seismographs that the Apollo astronauts left behind that the moon is almost geologically dead. The equipment did detect a few faint and infrequent moon quakes emanating about 700 kilometers below the lunar surface. But these were far too weak to rearrange the surface enough to be seen from the earth. Today we have a pretty solid understanding of the moon's geology. It seems that only a large meteor impact could alter the moon's appearance, and those happen too rarely to be worth looking for.

But in the 1950s astronomers were still wrestling with these issues, and "The Amateur Scientist" was right in the thick of things. For the student of scientific methods, these articles make for interesting reading. Not only do they provide insights into the controversies within lunar science as it existed at the time, but they also illustrate how good the human mind can be at seeing just what it wants to see. Doing science is all about not fooling yourself. Good scientists studiously practice self-skepticism, where as poor scientists fool themselves often. Only the good scientists consistently make contributions to human understanding.

Ed.

observer, who always wishes to see something, succeeds in representing fleeting detail. In the region of Eratosthenes it has seemed to me that a great number of lines have their origin in the contrast between two bright areas; when two bright regions form, one can see between them a dark line. And the optical theory explains the displacements of the fields and canals by a change in the maximum distribution of elementary spots scattered over a region."

17 CURIOUS AMATEUR OBSERVATIONS OF THE MOON

Conducted by Albert G. Ingalls, January 1953

Astronomers define selenography as the study of the surface of the moon. They are so busy with the stars and the universe, however, that they have no time for selenography. Thus the moon has long been left almost entirely to advanced amateur astronomers, who find it made to order for their more limited resources—and also endlessly fascinating. Of the more than 1.3 million craters on its surface, some 30,000 are easily visible through a telescope.

When telescope users cease to look at the moon in a merely desultory manner and begin to observe it, they have begun to be selenographers. They single out small areas or formations and study them minutely throughout the long lunar day, at every hour of which the changing angle of the sunlight alters their appearance. While doing this, they learn the lunar map, partly by copying and then drawing it from memory, partly by making sketches directly at the eyepiece of the telescope or photographing the lunar surface. If these pictures are dated and saved from the very beginning, these observations will be all the sharper. The ability of the eye to see detail improves immensely with practice.

The training of the powers of observation would be tedious if telescope owners did not fan their interest by reading the literature to learn what other selenographers are doing. Perhaps the easiest way to start is by joining the Association of Lunar and Planetary Observers (ALPO, *www.lpl.arizona.edu/alpo/*) and reading its monthly journal *The Strolling*

135

Astronomer. By following up the leads in its articles, beginners soon learn their way about in the world of selenography and are introduced to the many controversies in this field.

Of these perhaps the most interesting is the question: Does the moon's surface ever change? Most astronomers believe that it does not, even in fine details. On the other hand, most selenographers believe that small changes do occur. Dismissing the theories of the astronomers on the ground that they do not observe the moon, the selenographers insist that their minute and systematic observations over many decades have confirmed a number of changes.

The most noted change was the disappearance of the crater Linné, six miles in diameter, some time between 1843 and 1866. After its disappearance, a white spot surrounding the crater steadily diminished in size and brightness until 1897, but then grew again and has now regained its earlier size. For many years the American selenographer William H. Pickering noted irregular changes in the sizes and shapes of dark regions within the ring plain Eratosthenes. (See Chapter 16, "Observing Changes on the Moon.") Then there is the ring plain Plato, which is obscured at irregular intervals by some kind of haze or vapor.

In 1942 Walter H. Haas (who went on to found ALPO—Mars Association of Lunar and Planetary Observers, *www.lpl.arizona.edu/alpo/*—in 1947) summarized changes that selenographers had seen in 21 lunar formations in a series of articles in *The Journal of the Royal Astronomical Society of Canada.* In 1952 the selenographer H. P. Wilkins, director of the lunar section of the British Astronomical Association and maker of the best detailed map of the moon, described 15 anomalies that he had observed in 40 years of lunar observation with telescopes from 3 to 15 inches in aperture. Wilkins said: "Things do happen and are continually happening on the moon." It is not a dead world.

Some astronomers treat these many claims more open-mindedly than they did 50 years ago, when the U.S. astronomer Simon Newcomb said dogmatically that the moon was a world on which nothing ever happens. Most open-minded is the textbook *Astronomy* by William T. Skilling and Robert S. Richardson. They describe the work of selenographers and note that "astronomers are extremely skeptical of changes on the moon, considering them to arise from differences of illumination, unsteadiness of the earth's atmosphere and the inherent difficulty of seeing and recording fine details which are just at the limit of visibility." But they concede that "most astronomers have not systematically studied the moon."

In the years 1951–52 readers of *The Strolling Astronomer* followed a long account (15,000 words) of a lunar formation which seemed to the Baltimore selenographer James C. Bartlett, Jr., to be playing a game of hide and seek. It was a puzzle that fascinated selenographers for over 100 years.

Some time prior to 1837 Johann Maedler of Berlin saw a remarkably perfect square on the region of the moon between the ring plain named Fontenelle and the walled enclosure called Birmingham. The square was 65 miles on a side and had walls one mile thick. It is shown in Roger Hayward's drawing below, reproduced from a drawing published in 1837 by Maedler and his collaborator Wilhelm Beer. On the floor of the square Maedler saw a very regular cross. In 1876 Edmund Neison, director of the Natal Observatory in South Africa, also published a lunar map containing Maedler's square, which he had observed himself with a six-inch refracting telescope. He described it as "a square with regularity and perfect form, its walls from 250 to 3,000 feet in height." Neison's drawing is reproduced here by Hayward, himself a selenographer.

Thus two of the founders of selenography testified that the square was real in their time. Yet in 1949 Bartlett, after studying the area of the square for more than an hour with his 3.5-inch reflecting telescope on a very fine

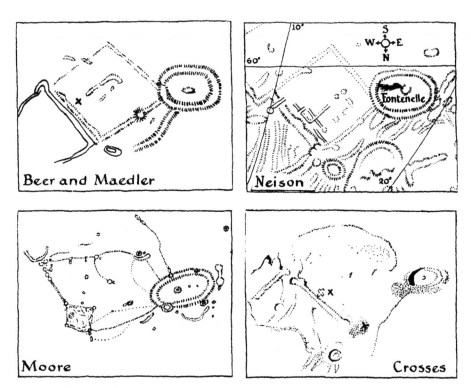

Figure 17.1 Drawings of the same lunar formation by four observers

night, failed to find it! The "fact emerged," he says, "with the impact of a hydrogen bomb. No such formation existed." Another selenographer, E. J. Reese, confirmed the disappearance. Bartlett later found two walls of the "square" (the ones on the lower left-hand and right-hand sides of the drawings), but he could detect no trace of the other sides.

Bartlett then examined old photographs of the moon made in Maedler's and Neison's time. (The first photograph of the moon was made in 1840 by the U.S. astronomer Henry Draper.) The primitive photographs were inconclusive, but in the first clear ones, made during the 1870s by the U.S. amateur astronomer L. M. Rutherford, the "square" did not appear. Bartlett concluded that what Maedler and Neison had seen with their telescopes actually was a square, which had "ceased to exist by 1874 and perhaps earlier."

The next participant in the discussion in *The Strolling Astronomer* was Patrick A. Moore, secretary of the lunar section of the British Astronomical Association. In July, 1951, he said that the Bartlett article had aroused a great deal of interest in Britain. He sketched the area with his 8.5-inch reflector, using powers from 200 to 400. Moore, a very experienced observer, found three walls of the square clearly visible and the fourth faintly so at times. But the square was not the neatly geometrical, fortress-like form Maedler had described, and all but one of its walls were very low.

Moore's findings are pictured in the drawing labeled with his name. The wall at the bottom of the drawing is extremely low. At its left end is a small quadrilateral bounded by four hills connected by low ridges. It contains a small crater and fine detail of the kind selenographers delight in delineating as a test of their observing skill. The left wall, extending upward from this corner, ends in a series of heights which are very conspicuous during the times of the month when the shadows reveal them. The wall at the top of the drawing is very low indeed and perhaps discontinuous. At its right-hand end is a prominent crater. Finally, on the fourth side of the square is a trace of a wall so low that it has to be caught under favorable conditions of illumination to be seen at all.

Have changes occurred since Maedler and Neison described the prominent, fortress-like walls they thought they had seen? Moore declares: "The evidence for change is totally inadequate." He excuses Maedler on the ground that he had only a small telescope (a 3.75-inch Fraunhofer refractor with magnification 300) and that it is human "to err sometimes." And he disqualifies Neison because the latter's map was made mainly from Maedler's. Supporting Bartlett's observations, Moore suggests that "it would be a fitting gesture to attach the name of Bartlett to the curious formation that has been referred to as Maedler's square."

Moore's careful observations seemed to have settled the question. But 14 months later Bartlett announced new evidence. Photographs taken with the 36-inch Lick refractor, he said, fail to show any eastern wall for the square. He suggested that the extremely low object seen there by Moore could not have been seen by Maedler and might possibly be the remains of a wall that was actually present in Maedler's time. He dismissed the idea that Maedler's observation was due to the inadequacy of his telescope, noting that Maedler was able to see objects much smaller than the square.

Later Bartlett rediscovered Maedler's cross—a dull, whitish formation—and Reese verified the find. "Now," said Bartlett, "this wonderfully

The history of lunar observations provides a great case study for all scientists, professional or amateur. As noted in the text, the British selenographer (moon mapper) Patrick A. Moore insisted that he could see mists appearing in some craters and strongly asserted that the mist was a real substance that filled the craters as they warmed in the sun's light. And Moore wasn't alone. A number of other observers likewise reported seeing these strange fogs. However, most lunar scientists at the time did not accept these reports because many observers couldn't find the fogs, no matter how often they looked. Further, mists on the moon didn't make sense. That's because ice, whether it's composed of water, or methane, or anything else, can evaporate through a process called sublimation. Ice on the lunar surface would find the vacuum of space and the heat of the sun to be an intolerable combination. Any surface ice that was warmed by direct sunlight would have sublimated away billions of years ago, with the individual molecules receiving enough solar energy to escape the moon's gravity completely. In fact, anything that produced gas on the moon when heated by the sun should have been dried out long ago because the gas molecules would escape. (Interestingly, some scientists still speculate about finding water ice on the moon, but only where the sun never shines, deep inside craters near the poles—and recent observations support this [see news release on page 140]. These locations cannot be observed from the earth.)

Moore was not dissuaded by these arguments. He and a cadre of other observers continued seeing fogs form in the lunar mornings. And, as stated in the text, the American selenographer W. H. Pickering went so far as to speculate that changes he observed were evidence of something moving over the surface, perhaps something that was alive.

We now know for certain that Moore's mists and Pickering's peccadilloes were phantoms of their own creation. Reconnaissance satellites orbiting

(continued)

the moon have never returned photos of fog-filled craters, and direct chemical studies of moon rocks reveal the lunar surface to be almost devoid of water or other volatile substances. And, of course, the moon has been proven to be far too hostile an environment for life to thrive.

In the end, the skeptics, like Italian astronomer Mentori Maggini, held sway. Scientific research is all about making sure you don't fool yourself. And Maggini knew how easy it is to be fooled, especially when the observer is anxious to make a big discovery. Experimentalists who work in laboratories often build into their procedures consistency checks to help ensure they don't fool themselves. But visual observers like the selenographers don't have that luxury; they are limited to only what they see. And that makes observing the moon, and other forms of purely visual observation, a game for only the most cautious and self-skeptical scientists.

So I invite you to make your own detailed observations of the moon. Not because you are likely to make an important discovery, but rather as a lesson in scientific humility.

Ed.

NASA Ames Research Center News Release

Latest Results from the Lunar Prospector Mission: More Polar Water Ice, Strong Local Magnetic Fields

The north and south poles of the Moon may contain up to six billion metric tons of water ice, a more than ten-fold increase over previous estimates, according to scientists working with data from NASA's Lunar Prospector mission.

Growing evidence now suggests that water ice deposits of relatively high concentration are trapped beneath the soil in the permanently shadowed craters of both lunar polar regions. The researchers believe that alternative explanations, such as concentrations of hydrogen from the solar wind, are unlikely.

Mission scientists also report the detection of strong, localized magnetic fields; delineation of new mass concentrations on the surface; and the mapping of the global distribution of major rock types, key resources and trace elements. In addition, there are strong suggestions that the Moon has a small, iron-rich core. The new findings are published in the Sept. 4, 1998 issue of Science magazine.

In March, mission scientists reported Prospector had found a minimum abundance of one percent by weight of water ice in the rocky lunar soil (regolith), corresponding to an estimated total of 300 million metric tons of ice at the Moon's poles. "We based those earlier, conscientiously conservative estimates on graphs of neutron spectrometer data, which showed distinctive dips over the lunar polar regions," said Dr. Alan Binder of the Lunar Research Institute, Gilroy, CA, the Lunar Prospector principal investigator. "This indicated significant hydrogen enrichment, a telltale signature of the presence of water ice.

"Subsequent analysis, combined with improved lunar models, shows conclusively that there is hydrogen at the Moon's poles," Binder said. "Though other explanations are possible, we interpret the data to mean that significant quantities of water ice are located in permanently shadowed craters in both lunar polar regions.

"The data do not tell us definitively the form of the water ice," Binder added. "However, if the main source is cometary impacts, as most scientists believe, our expectation is that we have areas at both poles with layers of near-pure water ice." In fact, the new analysis "indicates the presence of discrete, confined, near-pure water ice deposits buried beneath as much as 18 inches (40 centimeters) of dry regolith, with the water signature being 15 percent stronger at the Moon's north pole than at the south."

How much water do scientists believe they have found? "It is difficult to develop a numerical estimate," said Dr. William Feldman, co-investigator and spectrometer specialist at the Department of Energy's Los Alamos National Laboratory, NM. "However, we calculate that each polar region may contain as much as three billion metric tons of water ice."

Feldman noted he had cautioned that earlier estimates "could be off by a factor of ten," due to the inadequacy of existing lunar models. The new estimate is well within reason, he added, since it is still "one to two orders of magnitude less than the amount of water predicted as possibly delivered to, and retained on, the Moon by comets," according to earlier projections by Dr. Jim Arnold of the University of California at San Diego.

In other results, data from Lunar Prospector's gamma ray spectrometer have been used to develop the first global maps of the Moon's elemental composition. The maps show large compositional variations of thorium, potassium and iron over the lunar surface, providing insights into the Moon's crust as it was formed. The distribution of thorium and potassium on the Moon's near side supports the idea that some portion of materials rich in these trace elements was scattered over a large area as a result of ejection by asteroid and comet impacts.

While its magnetic field is relatively weak and not global in nature like those of most planets, the Moon does contain magnetized rocks on its upper surface, according to data from Lunar Prospector's magnetometer and electron reflectometer. The resulting strong, local magnetic fields create the two smallest known magnetospheres in the Solar System.

These mini-magnetospheres are located diametrically opposite to large impact basins on the lunar surface, leading scientists to conclude that the magnetic regions formed as the result of these titanic impacts. One theory is

(continued)

that these impacts produced a cloud of electrically charged gas that expanded around the Moon in about five minutes, compressing and amplifying the pre-existing, primitive ambient magnetic field on the opposite side. This field was then "frozen" into the surface crust and retained as the Moon's then-molten core solidified and the global field vanished.

Using data from Prospector's doppler gravity experiment, scientists have developed the first precise gravity map of the entire lunar surface. In the process, they have discovered seven previously unknown mass concentrations, lava-filled craters on the lunar surface known to cause gravitational anomalies. Three are located on the Moon's near side and four on its far side. This new, high-quality information will help engineers determine the long-term, altitude-related behavior of lunar-orbiting spacecraft, and more accurately assess fuel needs for possible future Moon missions.

Finally, Lunar Prospector data suggests that the Moon has a small, iron-rich core approximately 186 miles (300 kilometers) in radius, which is toward the smaller end of the range predicted by most current theories. "This theory seems to best fit the available data and models, but it is not a unique fit," cautioned Binder. "We will be able to say much more about this when we get magnetic data related to core size later in the mission." Ultimately, a precise figure for the core size will help constrain models of how the Moon originally formed.

Lunar Prospector was launched on Jan. 6, 1998, aboard a Lockheed Martin Athena 2 solid-fuel rocket and entered lunar orbit on Jan. 11. After a one-year primary mission orbiting the Moon at a height of approximately 63 miles (100 kilometers), mission controllers plan to the lower the spacecraft's orbit substantially to obtain detailed measurements. The $63 million mission is managed by NASA's Ames Research Center, Moffett Field, CA.

Further information about Lunar Prospector, its science data return, and relevant charts and graphics can be found on the project website at *http://lunar.arc.nasa.gov.*

• •

establishes Maedler's accuracy. Have we any further reason to doubt that Maedler had faithfully depicted the square?" A few months later, with Maedler's cross plainly visible, Bartlett observed to the east of it a smaller, dark gray cross very difficult to see. Neison discovered this cross long ago when Maedler had missed it—proving, Bartlett says, that Neison did not merely rehash Maedler's book for his own.

Is Bartlett's comeback on the accuracy of the early selenographers a proof? The most he asks is that the square be closely watched in the future. Meanwhile Moore and Wilkins have made a change in Maedler's square. On the great Wilkins map of the moon they have renamed it "Bartlett."

18 A PAUPER'S GUIDE TO MEASURING LATITUDE

Conducted by Albert G. Ingalls, March 1952

Astronomy connects people to the cosmos, sometimes literally. Today, any-one with access to a Geo Positioning Satellite (GPS) receiver can pinpoint their position anywhere on earth to within a few tens of meters. However, in the 1950's when the next two articles appeared, precise geolocation was a challenging problem. The first article explores how accurately an observer's latitude can be determined using extremely simple apparatus. The second article provides a fascinating historical perspective on this fundamental prob-lem and it gives a clever astronomical solution to determine both latitude and longitude. Further, it challenges you to develop your skill at making precision measurements. I suggest you begin by using the sun to determine your lati-tude (being careful, of course, never to look directly at the sun). Then try the lunar occultation method to accurately determine both the latitude and longi-tude of your telescope. Finally, compare these results with each other, and with a GPS receiver. I can think of no better way for amateurs to develop a sense of their own connectedness to the Universe at large than to use the sun, the stars and the moon to precisely determine something as personal as where they stand on earth.

Ed.

In 1913 William Brooks Cabot and Russell W. Porter together explored the St. Augustine River in Labrador. In later years they discussed the question whether explorers could determine the latitude in the field with-out a precision instrument. Porter tried it and described the experiment in the journal *Popular Astronomy*.

Cabot, he wrote, "deplored even the addition of a pocket sextant and horizon to a camper's pack where dead weights must be shaved to a minimum, and argued that the tools irreducible to the explorer—a knife, hatchet and fish line—together with Nature's available materials, would suffice to obtain latitude within a mile" by measuring the sun's altitude. Porter doubted this, but made the experiment at Springfield, Vt., and was greatly surprised to find that the specified precision could be reached if he used a steel tape instead of a stretching fish-line. By various refinements he found the latitude within a fifth of a mile. The method should interest explorers, Robinson Crusoes, escaped prisoners, amateur astronomers and others with intellectual curiosity.

Roger Hayward's drawings on the opposite page describe the simple principle of the method. At *C* in the lower right-hand corner is a nail driven into a small log exactly opposite a chosen mark on the plumb line *BC*. The distance *BC* is known, and attached to the nail at *C* is a steel tape with a sliding sight. Always keeping *BAC* a right angle, the observer sights the sun at its culmination or highest angle (local apparent noon) and measures *AC*. In the right triangle *ABC* the distance *BC* divided by the distance *AC* will give a decimal fraction which is the sine of the angle *ABC*. The same decimal is then found in a table of sines, [or these days, using the 'asin' function on a calculator. Ed.] and on certain dates in March and September when the sun is over the Equator the angle shown opposite will be the observer's latitude. On all other dates, corrections given in tables in the ephemeris of the sun must be added or subtracted. In either case a small correction, given by Porter as about one minute of arc in summer and two in winter, must be added for atmospheric refraction.

For convenience in calculation Porter used a tape with feet divided into tenths and hundredths, and interpolated to thousandths of a foot. A metric tape is just as convenient, and Hayward points out that an ordinary tape with 96 subdivisions per foot may not introduce significant error. An error of one thousandth of a foot affects the result about half a mile in latitude.

For a sight Porter used a tiny hole in the sliding sight shown in the drawing, covered with colored glass to protect his eye. [Use No. 14 welder's glass, which you can obtain at any welder's supply store. Ed.] He found it possible to bisect the sun's disk with the nail at *B* reliably within less than one minute of arc.

In a test of the method Porter made six sights between 11:41 a.m. and 12:08 p.m. on the same day, obtaining latitudes for Springfield varying from 43 degrees 17.2 minutes to 43 degrees 19.8 minutes. The mean of the six sights was 43 degrees 18.5 minutes. As a check he next determined the latitude with his theodolite and found it 43 degrees 18.3 minutes. Thus he

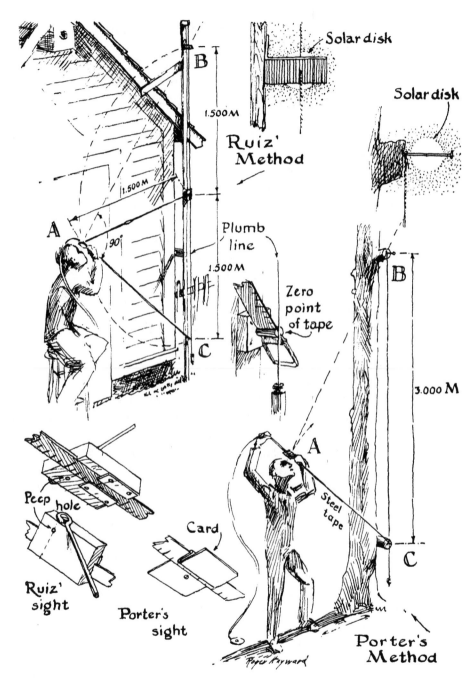

Figure 18.1 Finding the latitude without a precision instrument

145

had determined his latitude within .2 minute, or only 1,200 feet, without an instrument.

Theoretically the tape is unnecessary. Fish lines at *AC* and *BC* could be measured with any arbitrary, unknown unit of length, such as a stick, provided the same unit was used on *AC* and *BC*, but precision would be difficult.

John J. Ruiz of Dannemora, N.Y., tried Porter's experiment with more refined accessories and a modification of his own. Exactly halfway between *B* and *C* he pivoted a rod and hooked it to the sight, as shown at the left in the same drawing. "If you remember your Euclid," he writes, "you will note that the angle *BAC* is always 90 degrees." (An angle inscribed in a semicircle is a right angle.) The purpose of the rod was to exclude a possible source of error in Porter's method: "The observer's head," Porter wrote, "must be moved and the target shifted until the sun is bisected with the minimum length of tape."

Ruiz used a slotted pivot in place of the lower nail, and a sight with a vernier and 1/25-inch peephole (No. 60 drill) protected by two thicknesses of deep-colored cellophane. [Don't use cellophane! We now know that it does not block wavelengths that are just outside the visible spectrum, but that can damage your eyesight nevertheless. However, No. 14 welder's glass protects against all harmful wavelengths. Ed.] To average out the accidental errors he too made six observations and, since each took time, and since Joshua was not present to make the sun stand still, these sights were made at intervals before and after apparent noon and then reduced to the meridian. He came out neither worse nor better than Porter, that is, within about 1,200 feet of the true latitude.

Since it is impossible to evaluate all the contributing factors in the two observers' experiments, with the hidden chance errors, it cannot be known without more meticulous repetition by a third person whether or not Porter had already exhausted the precision inherent in the method. If he had, added accuracy in technique might bring only fictitious improvement.

The ephemeris of the sun is in the *Observer's Handbook* of the Royal Astronomical Society of Canada. [You can obtain a solar ephemeris online. Check out NASA's Jet Propulsion Laboratory's Horizon site at *ssd.jpl.nasa.gov/horizons.html*. Ed.] Ruiz provides refraction corrections which he says "are good enough for you and me." For 10-degree altitude of sun, add 5 minutes angle; for 20 degrees, 2.4 minutes; for 30 add 1.5; for 40 add 1; for 50 add .7; for 60 add .5; for 70 add .3; for 80 add .2, and for 90 add nothing.

19 PRECISION GEOLOCATION USING LUNAR OCCULTATION

Conducted by Albert G. Ingalls, January 1955

Where, precisely, am I? This is one of those easy-to-ask, impossible-to-answer questions. You must settle for an approximation. If you ask it while touring U.S. 80 from Plaster City, Calif., to Los Angeles, you may be content with the knowledge that you are less than a mile from Coyote Wells. But if you are an amateur astronomer setting up a telescope at the same site, you would prefer map information to the effect that you are at Latitude 32° 44′ 01″.29 North; Longitude 116° 45′ 24″.00 West.

Not even this seemingly precise pinpointing, however, would satisfy Colonel J. D. Abell and his associates in the Army Map Service. New methods of navigation, such as Loran, have disclosed gross errors in cartographic data. Particularly inaccurate are the positions of the oceanic islands; some important atolls in the Pacific appear to be as much as half a mile or more off their true positions on the map.

The personnel of Colonel Abell's bureau, in conjunction with the 30th Engineer Group under Colonel William C. Holley, developed an ingenious method of surveying by astronomical occultations which promises greatly improved accuracy. They invited amateurs—in or out of military service—to join in their fascinating research program.

"Our trouble," writes John A. O'Keefe, chief of the Research and Analysis Branch of the Army Map Service, "stems from the fact that we don't know straight up! If we had some way of pinpointing our zenith we could draw maps to any desired accuracy." In other words, if accurately known positions on the earth were correlated with one another by locating them with reference to the known positions of stars when they are at the zenith, the correlations would enable the cartographers to draw a good map of the world.

In principle the job is simple. You wait until a selected star of known position is directly overhead and clock it. Accurate timing is necessary because the relationship of the earth's surface to the sky changes continually as the earth rotates. Time signals broadcast from the U.S. Bureau of Standards' station WWV make precise clocking easy.

The usual instrument used for locating the zenith is a transit, which relies on a plumb bob or its counterpart, the bubble level. The source of error resides right here in these two gadgets, according to O'Keefe. Both the plumb bob and the bubble are thrown out of true by local irregularities in the density of the earth's crust which distort the gravitational field. Attempts have been made to correct for local deviations, but "this sort of guesswork gets you nowhere," says Floyd W. Hough, chief of the Service's Geodetic Division. "Even if you could estimate the effect of surface features accurately, you still would need information about conditions underground. Density varies there, too, and generally in the opposite direction."

The Army men decided to fix positions on the earth by timing occultations of stars by the moon as it moves across their positions in the heavens. One method of using the moon as a geodetic instrument is to photograph its position in relation to stars in the background at a given instant; it has been possible in this way to get fixes accurate to a tenth of a second of arc, which means locating positions on the earth with an accuracy within 600 feet. However, considering that this distance is more than twice the width of an aircraft runway, the desirability of still greater accuracy is obvious. The Army Map Service set out to improve on the accuracy of fixes by the moon's occultations.

The best telescopes, such as the 200-inch reflector on Palomar Mountain and the largest refractors, have a theoretical resolving power considerably better than .1 second of arc. But you cannot carry them from place to place on the earth, and furthermore their resolving power has practical limits, imposed by poor seeing conditions, distortion of the optical train by variations of temperature and so on. Above all there is diffraction, the master image-fuzzer, which arises from the wave character of light itself. Because adjacent waves interfere with one another, the light from a distant star does not cast a knife-sharp shadow when it passes the edge of the moon. Waves of starlight grazing the moon's edge interact, diverge and arrive at the earth's surface as a series of dark and light bands bordering the moon's shadow. The first band, the most pronounced, is about 40 feet wide.

The solution hit upon by O'Keefe and his associates was a new way to use a telescope which makes it capable of incredible resolution. They developed a portable rig (which amateurs can build) that plots lunar positions to within .005 second of arc as a matter of everyday field routine—resolution equivalent to that of an 800-inch telescope working under ideal conditions! It can also do a lot of other interesting things, such as measuring directly the diameters of many stars. It can split into double stars images which the big refractors show as single points of light. Some observers believe that it could even explore the atmosphere of a star layer by layer, as though it were dissecting a gaseous onion. Of greatest interest to the Army, the method measures earth distances of thousands of miles with a margin of uncertainty of no more than 150 feet!

The telescope that yields these impressive results has a physical aperture of only 12 inches. The design—a Cassegrain supported in a Springfield mounting—follows plans laid down by the late Russell W. Porter, for many years one of the world's leading amateur telescope makers.

The secret of the instrument's high resolving power is in the way it is used rather than in uniqueness of optical design. The telescope is trained on a selected star lying in the moon's orbit and is guided carefully until the advancing edge of the moon overtakes and begins to cover the star. Depending on the diameter and distance of the star, it may take up to .125 of a second for the moon to cover (occult) it completely. During this interval the edge of the moon becomes, in effect, part of the telescope—like a pinhole objective with an equivalent focal length of 240,000 miles. As the edge of the moon passes across the star, the intensity of the starlight diminishes, and the differences in intensity at successive instants are measured. It is as if a 240,000-mile-long tube were equipped at the distant end with a series of slit objectives—with the moon covering one slit at a time. The resolving power depends upon the great focal length.

The tiny successive steps in the starlight's decay are detected by a photomultiplier tube and a high-speed recorder. In principle the measurement of terrestrial distances by lunar occultation resembles measuring by the solar eclipse technique. The moon's shadow races over the earth's surface at about 1,800 feet per second. Except for differences in instrumentation and the mathematical reduction of results, the eclipse of the star is essentially the same kind of event as the eclipse of the sun. The insensitivity of the eye prevents star eclipses from making newspaper headlines, but photomultiplier tubes respond to such an eclipse strongly. They also detect the fuzziness caused by diffraction at the edge of the moon's shadow. The most prominent diffraction band, as previously mentioned, is some 40 feet across—the limit to which measurements by occultation are carried. The sharpest drop in starlight registered on typical recordings spans .015 second of time. Since the moon near the meridian has an average apparent speed of about .33 of a second of angular arc per second of time, the recorded interval of .015 of a second corresponds to .005 of a second of arc. This is the instrument's effective resolving power.

Any amateur who owns a Springfield mounting [*see page 151*] equipped with a high-quality mirror of eight inches aperture or larger can convert for high resolution work at a cost which is modest in proportion to the gain in performance. What one needs is a photomultiplier tube, a power supply, an amplifier and a high-speed recorder.

The photocell can be purchased for under $100. [The Hamamatsu corporation offers a wide range of options in photomultiplier tubes. Surf over to usa.hamamatsu.com for the latest details. Ed.] The amplifier must be of the direct-current type with a linear response good to at least 200 pulses per second. The recorder should be a double-channel job—one pen for registering time signals and the other for starlight. [These days pen recorders have been replaced by personal computers connected to analog to digital (AtoD) converters. Some of the least expensive and high-quality interfaces are sold by Vernier Software of Portland, OR. Set your browser to *www.vernier.com* to review the current options. Top of the line systems can be purchased from National Instruments in Austin, TX. You can find them at *www.ni.com*. Ed.] One also needs a filter to cut out the 400- and 600-cycle tone of WWV, so beloved of musicians. These units are available through dealers in radio equipment.

The eyepiece must be equipped with a cell for the photomultiplier tube and with a pinhole aperture for screening out unwanted moonlight. The pinhole (about .010 of an inch in diameter) is made in a metal mirror assembled in the eyepiece tube at an angle of 45 degrees, as shown in the drawing on page 152. A Ramsden eyepiece focuses on the pinhole. In operation the mirror is seen as a bright field with a small black speck, the pinhole, in the center. The star's image appears against the mirror as a brighter speck on the bright field. Thus it is easy to guide the image into position over the pinhole. When properly centered, some starlight strikes the edge of the pinhole, forming a small brilliant ring surrounding a jet-black speck. The ring aids in subsequent guiding.

Occultation observing has attracted a substantial following among amateurs. In the U.S. their interest in the work has been stimulated by the Occultation Section of the American Association of Variable Star Observers in Cambridge, Mass. The results of occultation observations

Figure 19.1　Springfield mounting equipped for photoelectric occultation work

have been used to establish irregularities in the rotation of the earth and to improve the tabulations of the moon's orbit.

Dirk Brouwer of the Yale University Observatory, who has made an exhaustive interpretation of the observations collected during the past century, sees an opportunity in the new photoelectric technique for the group to make an impressive addition to its already substantial scientific contribution. The photoelectric cell betters the response time of the eye (estimated at about .1 second of arc) by 100-fold or more and eliminates human variables. Thus it makes possible far higher accuracy in timing occultations. Moreover, the high-resolution aspect of the technique opens a whole new and relatively unexplored field for original work by amateurs. Star occultations, like solar eclipses, can be observed only in certain regions at particular times. A world-wide network of amateur observatories equipped for high resolution work could cover many more star occultations in any year than are accessible to the great telescopes of Southern California.

One serious drawback that prevents utilizing the full potential of the

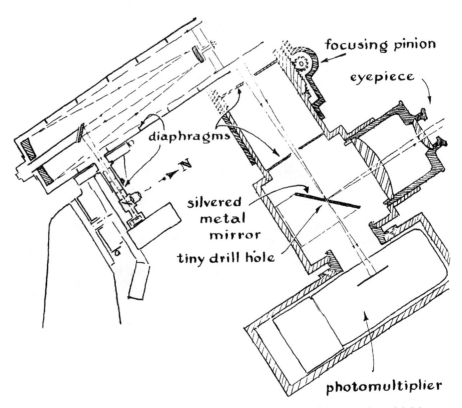

Figure 19.2   Optical path of the system (*left*) and details of the eyepiece (*right*)

increased accuracy at present is the irregularity of the moon's surface. If these irregularities are not allowed for in the calculations, the resulting position of the moon will frequently be off by several tenths of a second of arc. And if the star happens to be occulted at a point on the moon's limb where a high peak or low valley is located, the result may be off in extreme cases by two seconds of arc. A new study of the irregularities of the moon's surface by C. B. Watts at the United States Naval Observatory in Washington, expected to be completed soon, should make it possible to correct for the deviations with an accuracy matching the sensitivity of the photoelectric technique. [Of course, extensive and highly detailed lunar maps are now available online. Turn your favorite search engine loose on "lunar maps" for the most current options. Ed.]

The drawing on page 153 shows a pair of typical curves, recording the occultations of a sixth magnitude star and an eighth magnitude one. Note the jaggedness of the fainter star's curve. This is due to "noise," a term borrowed from radio and telephone engineering to describe random fluctuations in the output current of an amplifying device. The output of noise increases when the volume or "gain" control of the amplifier is turned up to compensate for a weak input signal. Noise originating in the photomultiplier (the principal source) can be reduced by chilling the tube with dry ice. The sharp drop in each curve marks the interval of occultation. Its steepness is determined principally by the diffraction pattern. In the case of some big stars, such as Antares, the effect of size can be seen in a flattening of the curve.

When a double-star system is occulted, the curve drops steeply for a time, indicating occultation of the first star, then levels off, and falls steeply again when the companion is occulted. The duration of the flat portion of the curve is the measure of the pair's separation. Some curves of Antares and other large stars show bends and twists which seem to come from bright and dark parts of the star's disk as well as from the stellar atmosphere. The proper interpretation of these records, however, is still considered an open question by some astronomers—another indication of the opportunity the technique presents to an amateur who enjoys original work.

"There is far better than an even chance that we shall stumble onto much that we didn't expect," writes O'Keefe. "We are examining stellar disks with greater resolving power than ever before. We shall certainly find a lot of close, fast binary stars. Perhaps we shall also find stars with extended atmospheres and all that. In occultations of very bright stars we are in a position to detect very faint, close companions. I really do not see how anyone getting into this sport can miss hooking some information of value, possibly even a really big fish."

In many respects photoelectric occultation seems almost

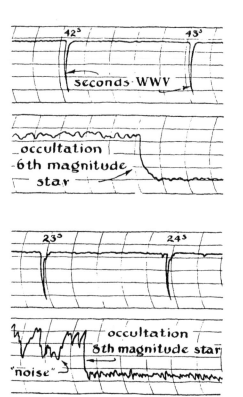

Figure 19.3 Recordings of star occultations

too good to be true. Neither poor seeing nor diffraction within the instrument has the slightest effect on the high resolving power of the method, and it is as precise when the moon occults a star low in the sky as overhead.

"The whole thing," writes O'Keefe, "no doubt gives the impression that a rabbit is being produced from a hat. It appears most surprising that such a powerful method for detailed examination of the sky should have gone unexplored for so long. This, of course, we enjoy. Our group did not invent the technique: It was suggested by K. Schwarzschild in Germany and A. E. Whitford in the U.S. It has not been exploited before because people simply could not believe that it works. But if I can get people to disbelieve thoroughly in something which is done before their eyes, then I have at least entertained them—and myself."

20 SUNDIAL POTPOURRI

Conducted by C. L. Stong, August 1959 and March 1964

Believe it or not, throughout The Amateur Scientist's 70+ year history, no topic has received more coverage than sundials. That's largely because the folks responsible for The Amateur Scientist's content for its first 50 years, Albert Ingalls and C. L. Stong, were both fascinated by these elegant devices. It's easy to understand why. Sundials are first-principle instruments; they connect the sun's instantaneous position in the sky with the precise local time. They are elegantly simple and have no moving parts, and yet they can be extremely sophisticated and amazingly accurate clocks. Some keep time to within 5 seconds. Moreover, sundials delight the mind; you can't grasp how they work unless you understand spherical geometry as well as the intricacies of the earth's orbit around the sun. I think this explains why modern-day sundial builders can spend months laboring over their creations, all the while wearing wrist watches that keep time to within thirty seconds per year.

The piece that follows combines two sundial designs offered by C. L. Stong. The first sundial provides a great illustration of the dynamics of the earth-sun system and explains why it is that the sun appears to move through our skies as it does. It would make a great teaching aid. The second would make a great student project. It is easy to build and illustrates all the basic principles of how more complicated sundials work.

Ed.

Why is it that a person who owns a perfectly good watch and several clocks will buy or build a sundial? It is not enough to say that a sundial makes a pleasant ornament in the garden. A deeper answer is that there is considerable intellectual charm in a device which, though it is motionless, converts the constantly changing motion of the sun into accurate time.

Of course a sundial that tells the time with any real accuracy is exceedingly rare. The problem is that the earth moves faster along its orbit in January than it does in July, and that the height of the sun's path across the sky changes every day. These difficulties account for the exceptional interest of a sundial constructed by Richard M. Sutton, professor of physics at the California Institute of Technology.

"If you leave a tennis ball undisturbed in the closet for a week," writes Sutton, "it turns completely around in space seven times! This simple fact, which ordinarily escapes notice, can be put to good use. Most people would say that the ball has not moved at all, yet they would admit the intellectual fact that the earth turns on its axis. The ball is turned by the earth around an axis parallel to that of the earth, and at just the rate at which the earth turns: 15 degrees per hour.

"If we combine the fact that the ball turns completely around once a day with an equally simple fact, we can convert any globe of the earth into a remarkable universal sundial that tells more about sunlight, the earth's motions in space and the conditions of sunlight in distant lands than might be supposed. The second fact is that the light falling on the earth from the sun comes in a flood of substantially parallel rays. Because of the great distance of the sun (some 93 million miles), even the extremes of a diameter of the earth are struck by rays that diverge by only .005 degree. This means that the angle subtended by a line 8,000 miles long seen at a distance of 93 million miles is about 1/200 of a degree. The significance of this fact will be apparent below.

"The rules for setting up the globe are simple and easily followed. It is rigidly oriented as an exact copy of the earth in space, with its polar axis parallel to the earth's axis, and with your own home town (or state) right 'on the top of the world' (where most of us like to think we belong anyway!). First turn the globe until its axis lies in your local meridian, in the true north and south plane that may be found by observing the shadow of a vertical object at local noon, by observing the pole star on a clear night, or by consulting a magnetic compass (if you know the local variation of the compass). Next turn the globe on its axis until the circle of longitude through your home locality lies in the meridian just found. Finally tilt the axis around an east-west horizontal line until your home town stands at the very top of the world. If you have followed these three steps, then your meridian circle (connecting the poles of your globe) will lie vertically in the north-south plane, and a line drawn from the center of the globe to your own local zenith will pass directly through your home spot on the map. Now lock the globe in this position and let the rotation of the earth do the rest. This takes patience, for in your eagerness to see all that the globe can tell you, you may be tempted to turn it at a rate greater than that of the turning of the earth. But it will take a year for the sun to tell you all it can before it begins to repeat its story.

"When you look at the globe sitting in this proper orientation—'rectified' and immobile—you will of course see half of it lighted by the sun and half of it in shadow. These are the very halves of the earth in light or darkness at that moment. An hour later the circle separating light from shadow has turned westward, its intersections with the Equator having moved 15 degrees to the west. On the side of the circle west of you, the sun is rising; on the side east of you, the sun is setting. You can 'count up the hours' along the Equator between your home meridian and the sunset line and estimate closely how many hours of sunlight still remain for you that day; or you can look to the west of you and see how soon the sun will rise, say,

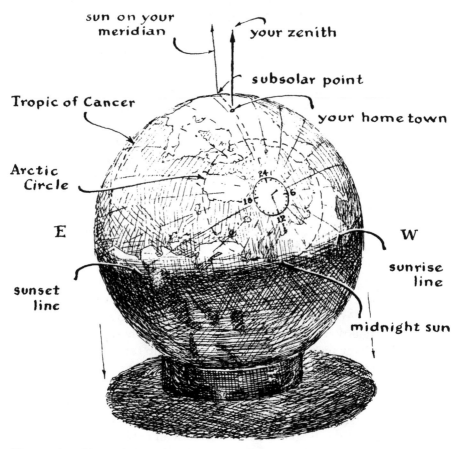

Figure 20.1 The global sundial, showing the North Pole on June 21

in Japan. As you watch the globe day after day, you will become aware of the slow turning of the circle northward or southward, depending upon the time of year.

"Let us take an imaginative look at the globe as it sits in the sun. Suppose it is during those days in June when the sun stands near the zenith in Hawaii. The globe dial shows that it is still sunlight at 9:30 p.m. in Iceland, that the midnight sun is shining on the North Cape of Norway. It is between late and early afternoon on the U.S. mainland, being about 6 p.m. in New York and 3 p.m. in San Francisco. The eastern half of South America is already in the darkness of its longer winter nights. The sun has recently risen *next day* in New Zealand and the eastern half of Australia, and most of China and Siberia are in early-morning light, whereas in Japan the sun is already four hours high in the sky. Alaska is enjoying the middle of a long summer day with the sun as high in the sky as it ever gets. Seattle is in early afternoon about two sun-hours ahead of Honolulu in the midst of a 16-hour day, while Sydney, Australia, is just starting a day with only 10 hours of sunlight.

"Now you don't have to be in Honolulu to see all this happening. Your own globe tells it to you. The same is true for persons who set up their globes in Fairbanks, Honolulu, Tokyo, Caracas, Havana or anywhere else. They will all see exactly the same story at the same time if in each place they have taken the small trouble to set up their globes for their own home towns as directed. If we choose, we can follow the progress of the circle of light and dark through the year. Three months later, for example, when the sun has returned close to the celestial equator, and when it passes day by day close to the zenith along our own Equator, we will see the circle between light and dark apparently hinged on the polar axis of our globe. This is the time of the equinoxes, when every spot on earth has 12 hours of light and 12 hours of darkness. On December 21 the sun will have gone to its position farthest south, now failing to light any spot within the Arctic Circle but lighting the region within the Antarctic Circle completely (as you may see by stooping and looking at the lower part of your globe).

"From its position farthest south, the sun starts its way north again at a rate that may seem painfully slow for those in northern latitudes who wait for spring. By March 21 it has again reached the Equator, and we find it at the vernal equinox, the astronomers' principal landmark. Through the centuries this was the time for the beginning of the year. Only as recently as 1752 did December cease to be the 10th month of the year, as its name implies. January 1, 1752, was the first time that the calendar year began in January in England and the American colonies! At the vernal equinox in March there is a sunrise lasting 24 hours at the North Pole, and a sunset

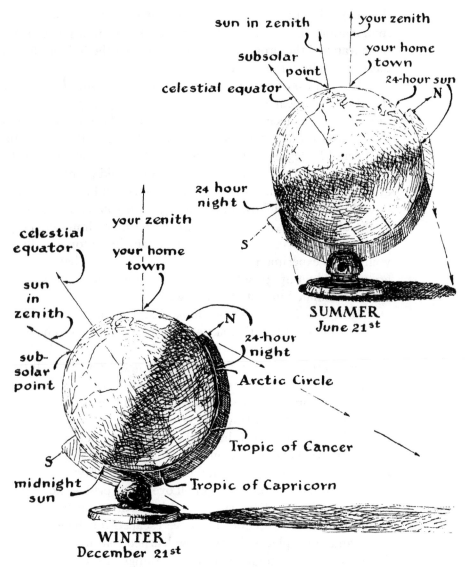

Figure 20.2 The illumination of the global sundial in winter and summer

lasting 24 hours at the South Pole. Now, as the months advance, we will find that on June 21 this circle of light has advanced to its position farthest north. Sunlight does not enter the Antarctic Circle on the bottom side of your globe at all, but it extends clear over the North Polar region to the Arctic Circle beyond. At noon in your garden on this day you will see how people living on the meridian 180 degrees from your home are enjoying the midnight sun, provided they live within the Arctic Circle. Thus in imagination we have made a complete trip around the earth's orbit and have watched the progress of sunlight during the 365 or 366 intervening days—all right in the garden.

"It is not easy to appreciate the fact that the sun's rays are parallel as they fall on the earth. Let me suggest a simple experiment. On a bright morning take a piece of pipe or a cardboard tube and point it at the sun so that it casts a small, ring-shaped shadow. Now if at the very same moment someone 120 degrees east of you—one third the way around the world—were to perform the same experiment, he would point his tube westward at the afternoon sun. Yet his tube and yours would necessarily be parallel to within a very small fraction of a degree. If you point the tube at the sun in the afternoon, and someone far to the west simultaneously does the same in his morning, his tube will again be automatically parallel to yours. This experiment will help explain how it is that, when our globes are properly set up, people all over the world who are in sunlight will see them illuminated in just the same way.

"How easy it is, with this global dial, to imagine oneself in a distant land, seeing the sun in that sky at that time of day. A pin held at any point on the globe immediately shows the direction of the shadow of a man standing at that spot. Your globe has become a 'terrella,' a little earth that shows what the big earth is doing in space.

"It was from long experimenting with a precision sundial drawn on the floor of my office at Haverford College that I slowly came to the idea of this dial. I had developed that dial to the point where I could tell the time within five seconds. But the global dial is more exciting. When it came to me, I was enthralled by its simplicity and profundity: to be able to see at a glance everything about sunlight all over the world without budging from my own garden or office. However, I had a strong feeling that an idea so simple and universal could not have escaped intelligent people at other times and other places. I have now learned that it was recognized some 300 years ago, when globes were playthings of the wealthy. People were then regarding their world with new understanding, made much richer by the great sailing explorations and the increasing recognition of the earth's sphericity. To be sure, the early Greeks had seen that the earth must be a sphere. For example, Archimedes based his great works on floating bodies on a proposition that reads: 'The free surface of any body of liquid at rest is part of a sphere whose center is the center of the earth.' Imagine that for 200 B.C.! There is much evidence in the writings of the Greek mathematicians that they appreciated this fact. Their estimates of the earth's size were correct in principle and not bad in actual result, but people seem to have ignored their observations and the reasoning behind them until the great age of exploration which we date from Columbus and the discovery of the New World.

"In a book on sundials by Joseph Moxon, first published in 1668, there is a description of 'the English globe, being a stabil and immobil one, performing what ordinary globes do, and much more.' Moxon, who was hydrographer to Charles II (and whose book was dedicated to Samuel Pepys, Principal Officer of the Navy), ascribes this globe to the Earl of Castlemaine. It seems certain that the globe existed in London by 1665. In 1756 another global sundial was described by Charles Leadbetter. Consider the delightful title of Leadbetter's book: 'MECHANICK DIALLING, or the New Art of Shadows, freed from the Obscurities, Superfluities, and Errors of former writers upon the Subject—the whole laid down after so plain a method that any person (tho' a Stranger to the Art) with a Pair of Compasses and Common Ruller only, may make a Dial upon any Plane for any place in the World, as well as those who have attained to the greatest Knowledge and Perfection in the Mathematics. A work not only usefull for Artificers but very entertaining for Gentlemen, and those Student at the Universities that would understand Dialling without the Fatigue of going through a Course of Mathematics.' They knew how to make full use of a title page in those days!

"Leadbetter tells how to erect an immobile stone sphere and inscribe a map on it. He says: 'According to their true latitudes and longitudes (for various spots on earth) you may discover any moment when the Sun shines upon the same, by the illuminated parts thereof, what Places on Earth are enlightened, and what Places are in darkness. . . . The Extremity of the Shadow shows likewise what Places the Sun is Rising or Setting at; and what Places have long Days; these with many more curious Problems are seen at one View, too many to be enumerated in this place. The dial is the most natural of all others because it resembles the Earth itself, and the exact manner of the Sun's shining thereon.' Leadbetter suggests that a pin be placed at each pole in order to use the global sundial to tell time. Around each pin are 24 marks—one every 15 degrees—corresponding to the hours; the time is read by noting the position of the pin's shadow with respect to the marks. I, too, have used this system. Leadbetter adds: 'As you see, that [pin] at the North Pole will give the hour in summer, that at the South Pole the hour in Winter.'

"There is *no* spot on earth with which we do not at some time during the year share the light from the sun. One might object that surely the nadir, that spot directly beneath our feet on the other side of the earth, has no sunlight while we ourselves have it; but atmospheric refraction keeps the sun in the sky longer than geometry alone predicts, making every sunrise about two minutes early and every sunset about two minutes late.

"It is easy to tell from the global sundial just how many hours of sunlight any latitude (including your own) will enjoy on any particular day. All you need to do is to count the number of 15-degree longitudinal divisions that lie within the lighted circle at the desired latitude. Thus at 40 degrees north latitude in summer the circle may cover 225 degrees of longitude along the 40th parallel, representing 15 divisions or 15 hours of sunlight. But in winter the circle may cover only 135 degrees, representing nine divisions or nine hours. As soon as the lighted circle passes beyond either pole, that pole has 24 hours of sunlight a day, and the opposite pole is in darkness.

"One or two other concepts may make the dial even more useful. First, we can answer the question: Where is the sun in the zenith right now? Can we find the spot on earth where men find their shadows right at their feet? Easily. Hold the end of a pencil at the surface of the globe and move it until its shadow is reduced to its own cross section. When the pencil points from the center of the sun to the center of the globe, the spot on the map where its end rests corresponds to the point immediately beneath the sun. Better still, if you use a small tube instead of a pencil, you can let the sunlight pass down the tube to cast a ring-shaped shadow at the point beneath the sun. This point is important, because it gives the latitude and longitude of the sun at that moment and locates the center of the great circle of daylight. Ninety degrees around the globe in any direction from that point is a point where the sun appears on the horizon, either rising or setting. At the north and south extremes of the circle between light and darkness are 'the points where sunrise and sunset meet.' In June, for example, the southern point shows where the sun barely rises and then promptly sets in Antarctica, and the northern point (beyond the North Pole) shows where the midnight sun dips to the horizon and immediately rises again.

"To find the point directly beneath the sun still more accurately, you can construct a simple cardboard tripod. Just cut three identical pieces of cardboard and fasten them together with gummed paper. When the tripod stands on a level table, the line joining the three vanes is vertical. When the tripod rests on the surface of your globe, the line extends outward along a radius of the globe. If you move the tripod about until the shadow of its three vanes disappears into three lines, you find the subsolar point at the junction of these lines. Once you have found the subsolar point, and hence the exact location of the sun in our system of coordinates, it is a simple matter to count off the hours since or until your local noon, to tell your local sun-time, to forecast the time until sunset and to tell how long it is since sunrise. You can also determine these things for locations other than your own.

"Perhaps this little sundial, so simply set up, will clarify the apparent motions of the sun, caused of course by the earth's daily rotation on its axis and its annual revolution around the sun. Surely it is fun to bring so much of the system of the world into your garden. The global dial can give one a fuller appreciation of the sunlight on which everyone depends. If it thereby strengthens your feeling of kinship for people far away, the instrument will have served you well."

Charles J. Merchant, a mathematician at the University of Arizona, submits the following description of a sundial for indicating standard clock time that can be made in less than an hour. "A sundial," writes Merchant, "even when it is perfectly constructed and correctly installed, generally indicates a time substantially different from standard clock time. This often leads people not familiar with the beautiful intricacies of sun time to the erroneous conclusion that a sundial is an inherently inaccurate device. Sundials that indicate time correctly to within one minute can be constructed with no great difficulty; with refinements they can be accurate to within a few seconds.

"The difficulties with sun time versus standard time stem from two sources. The eccentricity of the earth's orbit and the obliquity of the ecliptic cause the sun to gain or lose as much as a minute a day over considerable periods of time, with accumulated inaccuracies of plus or minus 15 minutes at certain times of the year. The correction for this variation is known as the equation of time. When this correction is applied to the reading of a sundial, the result is local mean time. Local mean time, however, is the same as standard time in the U.S. only in those cities whose longitude is 75, 90, 105 and 120 degrees. In all other localities standard time differs from local mean time by a constant amount depending on the longitude of the place. This second cause of a sundial's apparent inaccuracy is known as the longitude correction.

"Two corrections must therefore be applied to the reading of a conventional sundial in order to derive standard time: the equation of time, which varies from day to day, and the longitude correction, which is constant for a given place.

"Numerous methods, some of considerable ingenuity, have been devised for making a sundial indicate standard time directly. My sundial accomplishes this by means of a circular computer. The face of an equatorial-type dial is rotated by various amounts depending on the setting of a pair of disks. When the device is properly adjusted, it indicates standard time correctly to within better than five minutes. It operates only during the spring and summer months, from the vernal equinox to the autumnal equinox; during the other six months of the year the sun lies below the equatorial plane. A set of disks could be calibrated for this interval, but they have not been included with this model.

"The dial was designed to be cut out and mounted on thin cardboard, using a nonwrinkling cement. Rubber cement does an excellent job, but it is not permanent. After they have been mounted on the stiff backing the parts are carefully cut out. The base is then cut off for the latitude of the place where it is to be used and bent at right angles along the broken lines. When properly mounted on a baseboard and placed on a level surface, the face makes an angle with the horizontal equal to the colatitude of the place.

"The disks are then assembled on the face. A needle, pushed through the center mark from below, serves both as the means of assembly and as the gnomon. The gnomon should be as exactly perpendicular to the face as possible!

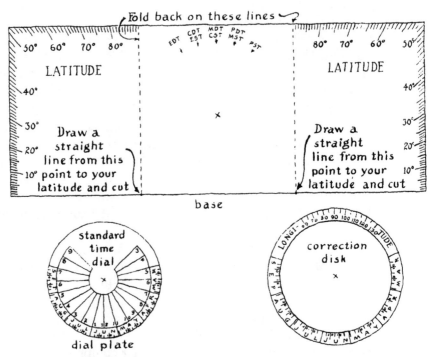

Figure 20.3 Details of a sundial that indicates standard time

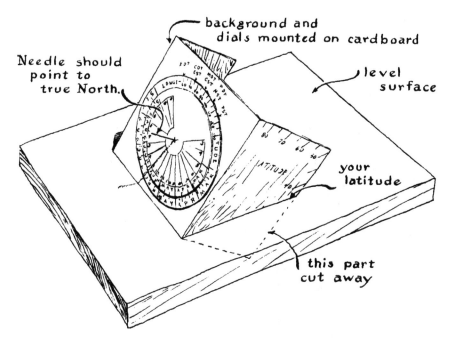

Figure 20.4 How the sundial is mounted

"The larger of the two disks, the correction disk, is placed face up on the needle first, then the smaller of the disks. Finally a small piece of cardboard, to act as a retaining washer, is pressed down on the needle.

"To operate the sundial the correction disk is first rotated so that the longitude of the place is opposite the arrow on the face that indicates the local time zone. (The abbreviations are self-explanatory: CST means Central Standard Time, MDT means Mountain Daylight Time, and so on.) For example, when the dial is to be used in New York City during the period of Eastern Daylight Time, the correction disk is turned so that 74 degrees, the longitude of New York City, is at the arrow marked EDT. This disk will require further adjustment. It can be fixed to the base plate with a small piece of drafting tape. The tape must not interfere with the movement of the smaller disk, however. This disk must be adjusted every few days.

"The correction disk carries a date scale that is graduated nonlinearly to correct for the equation of time. The outer edge of the dial plate also carries a date scale, but this scale is graduated linearly. When the dial plate is rotated so that a given date on the dial plate coincides with the same date on the correction disk, the time scale is automatically rotated by the amount necessary to correct for the equation of time on that date.

"Finally, the dial is set on a level surface—in the sun—with the gnomon pointing exactly north. The dial then indicates correct standard or daylight time. For instance, assume that the dial is to be used in New York City on July 10. The latitude of New York City is 41 degrees. The base support should be cut for this angle. The longitude of New York City is 74 degrees. On July 10 Eastern Daylight Time is in effect. The correction disk is therefore rotated so that longitude 74 degrees is at the EDT arrow. The correction disk should be taped to the face plate. The smaller disk is then turned so that July 10 coincides with the July 10 date on the larger disk. The dial is next placed on a level surface in the sun with the gnomon pointing exactly north. The shadow of the gnomon now indicates correct Eastern Daylight Time for New York City.

"It should be noted that this dial will indicate correctly even in those cases where a city operates under an incorrect standard time zone. Certain cities in eastern Indiana and western Ohio, for example, operate on Eastern Standard and Eastern Daylight Time even though they are well within the Central Standard Time zone. In these cities follow the rule of setting the longitude of the city to the arrow representing the time zone under which it operates. The dial will indicate the correct clock time.

"Although the dial is calibrated only for the time zones of the continental U.S., it can be used anywhere in the Northern Hemisphere. Merely add or subtract from the longitude of the place that multiple of 15 degrees which results in a longitude within plus or minus 7.5 degrees of 90 degrees, and then use the resulting longitude with the arrow for CST or CDT, depending on whether standard or daylight time is in use. All longitudes must be converted to longitude west of Greenwich, however. Thus longitudes east of Greenwich should be subtracted from 360 degrees to give west longitude."

21 HOW TO STUDY ARTIFICIAL SATELLITES

Conducted by C. L. Stong, January 1958

In the late 1950's and early 1960's the public was fascinated with artificial satellites. Back then the United States was losing the space race, while the Soviet Union moved from triumph to triumph. So the occasional points of light streaking through the heavens provided an uneasy public with a constant reminder of the precarious and shifting balance of power that threatened our very survival in the dawning nuclear age. It was in this climate of fear that "The Amateur Scientist" began exploring ways of observing these new additions to the night sky. In this, I believe that C. L. Stong did the free world a great service.

Many young people took up Stong's challenge and enthusiastically went into their backyards to analyze the orbits of passing satellites. I was one of them. For me, observing these objects and deducing the details about their trajectories demystified them. The hours I spent on my patio measuring their transit times and finding their angular speeds helped me understand the potential of these devices, both good and bad. And I developed an immense respect for the American and Soviet engineers who had mastered the art of placing them so precisely in their orbits. Somehow, studying satellites made me feel a kinship with the people who created them, and that made me much less anxious about having these machines in orbit.

Of course, technology has advanced tremendously since these methods were first published. Today's amateurs can simultaneously track dozens of individual satellites from the comfort of their own dens using inexpensive commercial equipment. Many hams routinely use satellites to send signals to distant continents. But the simple first-principle techniques presented here can still provide an important visceral link between the observer and our species' accomplishments in space. If you love the night sky and appreciate the prowess of technology, or are just looking for a delightful challenge, I urge you to give these techniques a try.

Ed.

Artificial satellites and their associated hardware have opened new horizons for the amateur scientist. The amateur does not need elaborate equipment to keep track of a satellite's path, or even to predict how long a satellite will stay on orbit. Anyone can do it. Those most interested will doubtless acquire at least a low-power telescope or a pair of binoculars. Others will be able to settle for a pair of pocket mirrors, a sheet of glass and some adhesive tape. The amateur who prefers to make observations with a minimum of equipment can get into business with nothing more than three wooden slats and a stop watch.

These techniques require that the observer be outdoors, often before sunrise. For the comfort-loving amateur there is another means of keeping track of satellites, at least so long as they transmit radio signals. Many ordinary radio receivers have built-in converters which enable them to pick up the short waves broadcast by satellites. Even those receivers without converters are easy to adapt to short-wave reception.

Ralph H. Lovberg and Louis C. Burkhardt, physicists at the Los Alamos Scientific Laboratory, have suggested a way in which the amateur can determine the height of a satellite as it passes overhead. Their method is based on the Doppler shift, the apparent decrease of pitch noted when a source of sound, such as a train whistle, rushes past the observer. Radio-equipped satellites are in effect whistling objects. Their radio transmitters radiate signals at predetermined frequencies. In the case of the first two satellites the frequencies were approximately 20 and 40 megacycles per second. To measure the distance between the observer and a satellite, the signal from the satellite is tuned in on a radio receiver and mixed with a signal generated by a local oscillator of somewhat higher or lower frequency. The difference frequency is then amplified, fed into a loudspeaker and converted into sound. For example, when a signal of 40 megacycles is mixed with a locally generated signal of 39.997 megacycles, the amplified difference frequency of 3,000 cycles will produce a whistling sound pitched about two octaves above middle C. The period during which the pitch changes may last more than five minutes. The distance of the satellite is determined by measuring the pitch of the sound at brief intervals during this period, and noting the time at which each tone is identified. The frequency of the sound at each interval can be estimated by comparing it with the notes of a piano or, preferably, by following the whistle down the scale with an accurately calibrated audio-frequency oscillator. Whenever the signal and comparison tone (piano or oscillator) coincide (are in zero beat), the corresponding time should be recorded as read from the second hand of a watch or clock.

"In the case of our measurements of the first satellite," write Lovberg and Burkhardt, "we used a conventional Hallicrafter SX-28 receiver at 40 megacycles. The local beat-oscillator was left off and a surplus BC-221 frequency meter was coupled loosely (by means of a twist of insulated wire) to the antenna lead. The signal generator is set at approximately 3,000 cycles below the satellite frequency when the 'little traveler' is first detected. This results in an audible tone in the receiver output. The tone is now fed into a loudspeaker together with the output of an audio oscillator. In a typical run one sets the audio oscillator to a tone lower than that of the satellite, say 2,500 cycles per second, and waits for the satellite tone to drop to the same value. The zero beat between the two tones is first heard as a fluttering sound which diminishes in frequency to a slow swelling and fading of the 2,500-cycle note. At the instant the tone becomes steady, one records the time as well as the frequency (2,500 cycles in this instance) and quickly shifts the audio-frequency oscillator to, say, 2,400 cycles, and waits again for the matching of the tones.

"The resulting table of frequencies and times is then plotted as a curve like the one shown on page 170. This indicates the passage of the first satellite over Los Alamos, N.M., on October 13. Next we draw a line tangent to the curve at its steepest point. The slope of the tangent line represents the number of cycles per second that the frequency changes per second and varies in proportion to the distance of the satellite. If we call this slope m, then the distance of the closest approach of the satellite is given by the simple formula d = fv²/cm. Here f equals the frequency of the satellite's transmitter (40 megacycles), v equals the velocity of the satellite in miles per second (about 4.9 miles), c is the speed of light in miles per second. The velocity v will vary somewhat from the 4.9-miles-per-second value, depending on such factors as the eccentricity of the satellite's orbit. The amateur may obtain fairly accurate velocity figures from agencies such as the Smithsonian Astrophysical Observatory. These figures are sometimes reported in the daily press."

It was by this method that Lovberg and Burkhardt learned that the first satellite was coming over Los Alamos at a height of about 170 miles at night and 260 miles during the day. The figures are in good agreement with those released by the Smithsonian Observatory.

An optical means of getting the same result is suggested by Walter Chestnut, a physicist at Brookhaven National Laboratory:

"If an object is in a stable, circular orbit around the earth, then the gravitational force of the earth pulls the object with a force which equals its centrifugal force. The relation may be expressed in mathematical terms in such a way that the object's altitude may be determined by a simple

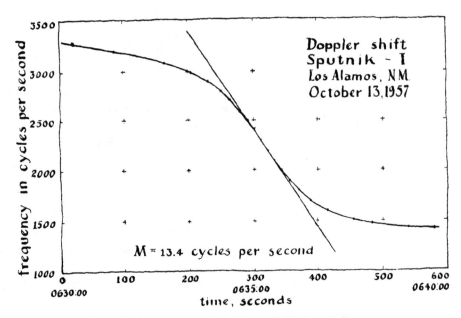

Figure 21.1 Curve for the determination of the height of a satellite

measurement of the number of degrees traversed by the satellite in one second as the object passes overhead. This angular velocity may be measured by timing the transit of the satellite as it passes stars, the positions of which are known. If the amateur does not have a star chart, the angle can be easily measured by a fixed astrolabe, a triangle of wooden slats nailed together. The astrolabe forms an angle of 10 degrees. The nails serve as sights; they should be painted white so that they will be clearly visible in poor illumination. The construction of the instrument should be as light as possible, because it must be swung into position quickly when a satellite is spotted. The two outer nails should be placed in line with the satellite's path and held steady. The number of seconds are then counted from the time the object appears to touch the first nail until it reaches the second one. The corresponding altitude in miles can then be read from the accompanying table [*see page 171*].

"Readers may calculate their own table for astrolabes of other angles by the equation:

$$d_1 = 282 \ t/a \ \sqrt{1 - .0705 \ t/a}$$

10 DEGREES TRANSIT TIME (SECONDS)	ALTITUDE (MILES)
2	56
3	84
4	112
5	138
6	166
7	193
8	219
9	246
10	273
12	325
14	375
16	426
18	477
20	626
25	650
30	770
35	890
40	1,000
50	1,225
60	1,450
70	1,660
80	1,860
90	2,060
100	2,250
120	2,620
140	2,980

Table 21.1   Table of transit times and altitudes

In this equation d_1 equals the altitude of the satellite in miles; t is the time of transit in seconds for an angle of a degrees. For an astrolabe of 15 degrees and a transit time of 60 seconds the altitude would be $(282 \times 60/15 \times (\sqrt{1 - .0705 \times 60/15})$, or $1,128 \times .8474$ (956) miles. The same calculation is carried out for each value of time desired in the table.

"Both the table and the equation assume that the satellite is observed within 15 degrees of the zenith. If the satellite is more than 15 degrees from the zenith, multiply the distance given in the table by the cosine of the angle between the orbit and the zenith. The equation and table also assume that the orbit is a circle. Most orbits, however, will be ellipses. But if one knows the maximum and minimum heights of the satellite, the table (and equation) can be corrected for ellipticity by the equation:

$$C = d_1 \left(\frac{d_e - d_1}{2(4,000 + d_e)} \right)$$

In this equation d_1 is the altitude obtained from the table (or the first equation). C is the distance (in miles) to be added to d_1 if C is positive, or to be subtracted from d_1 if C is negative. The average of the maximum and minimum altitudes of the satellite is represented by d_e. In the case of Sputnik I the maximum and minimum altitudes were respectively 570 and 170 miles; d_e accordingly equals 370 miles.

"By following these instructions carefully the amateur can determine the altitude of a satellite with an accuracy of about 20 miles for every 1,000 miles of its height. The limits of error will doubtless be determined more by the accuracy of the observer's measurements than by errors of the method.

"We have observed the rocket of Sputnik I on two occasions. The first time we merely wanted to find out if it was really there. On the second look the astrolabe was propped against a convenient tree. As the rocket traveled from one sighting nail of the astrolabe to the other, a transit time of 11.6 seconds was recorded. Unfortunately the tree swayed a little in the stiff morning breeze; this doubtless introduced an error of a few miles. The altitude for an 11.6-second transit, as it is read from the table, is 314 miles."

What about methods of measuring changes in the time it takes a satellite to make a trip around the world? As a satellite encounters atoms and molecules of the rarefied air at altitudes above 100 miles, it gradually loses speed. Paradoxically it appears to gain speed, because as it slows down it spirals closer to the earth and takes less time to complete its orbit. When the orbital time of a satellite decreases to about 87 minutes, the satellite will soon be consumed by friction with the lower atmosphere. Thus by timing the passage of a satellite during a few transits, its lifetime can be predicted. If the measurements can be made with good accuracy, two timings are sufficient for an approximate prediction.

A convenient instrument for timing the orbit of a satellite is the dipleidoscope, a device invented about 1860 by an English barrister named J. M. Bloxam. It consists of a pair of mirrors tilted toward each other at an angle of somewhat more than 90 degrees and covered by a sheet of glass as depicted on page 173. The three elements may be supported by a pair of end plates, as shown, or simply taped securely. Ideally the mirrors should

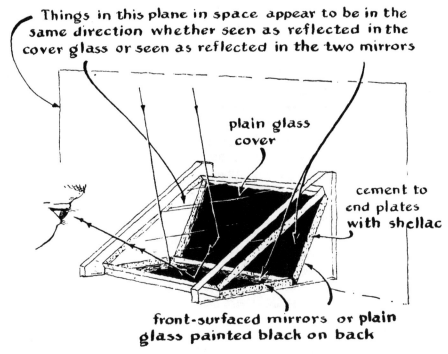

Things in this plane in space appear to be in the same direction whether seen as reflected in the cover glass or seen as reflected in the two mirrors

plain glass cover

cement to end plates with shellac

front-surfaced mirrors or plain glass painted black on back

Figure 21.2 Dipleidoscope to measure the time required for a satellite to go around the earth

be front-surface silvered and the cover glass should be silvered so that it reflects 38 per cent of the light striking it and passes the rest. But ordinary back-silvered mirrors (or even plain glass with a back-coating of black paint) will work, and the cover glass need not be silvered at all.

When the dipleidoscope is held at an angle which reflects light from the satellite into the eye, two images will be seen. One image is reflected by the cover glass, the other by the mirrors. As the satellite passes overhead, the two images move toward one another, merge and then pass out of the field of view in opposite directions. The time is recorded at the instant the images merge. The device should be set up in advance of the transit on a firm but easily adjusted support such as the ball-and-socket tripod head popular with photographers. The axis of the dipleidoscope should be adjusted roughly at a right angle to the anticipated path of the satellite. During one transit the instrument is adjusted so that the two images will merge. On the same transit the time is recorded. The instrument is then left undisturbed for a second observation during the next transit. In the case of Sputnik I observations made 24 hours apart would show the approximate apparent gain in time for 15 revolutions.

This article first appeared in 1958, and in the intervening decades satellite technology has become an important underpinning to our way of life. Satellites transmit many of your long distance calls. Hundreds of thousands of consumers watch television signals that have been down-linked from space, and find their way with Global Positioning System receivers. Some trendsetters carry pocket telephones that can link any two points on earth, compliments of satellites overhead. And some computer users now surf the Internet though satellite modems that could soon provide data throughputs of 38 megabytes per second.

Today, anyone equipped with some modestly priced radio equipment can now track, access and control dozens of these high-tech wonders to further their own communication interests. To find out more about tracking satellites by receiving their radio signals directly, check out *The ARRL Satellite Anthology* (Amateur Radio Relay League), or get their video tape *Getting Started in Amateur Satellites.* And there's a great resource online. Go to *www.spacelink.nasa.gov* to find NASA's SpaceLink page. Then search that site for "satellite tracking." NASA's page provides a great deal of information about satellites and how to monitor them.

With so much high-technology available, the stopwatch and nails approach to satellite tracking may seem quite crude. But relying on high-technology often obscures the basic physical principles that underlie what's going on. I like the methods described here because they get at the fundamental physics behind satellite motion.

In particular, observing the Doppler shift as a satellite transits overhead can be a delightful and highly instructive thing to do. Unfortunately, Stong's article omits some important details that are necessary to put it into practice. This amendment will point you in the right direction, but success will require experience in radio electronics.

These days, some well-heeled hams can directly observe radio Doppler shifts by amplifying a band of frequencies containing a satellite's signal and dumping the result directly into a spectrum analyzer, which displays the power detected at each frequency. These spectrum analyzers are extremely useful and can be found on the surplus market. Nose around your local electronics surplus shops, or set your browser to *www.ebay.com* and search for "spectrum analyzer." Experienced radio engineers can even hobble together their own systems for a few hundred dollars and a week's work. If you're interested, check out *www.amazon.com* for what's current.

Diehards may want to try Lovberg and Burkhardt's method. Their technique is a special application of superheterodyning, a cornerstone of radio communications that was old even when they wrote about it. For a detailed and extremely lucid discussion see *The Art of Electronics*, by Horowitz and Hill, Cambridge University Press.

Here's the quick explanation. A superheterodyne receiver "mixes" a steady frequency with the signal frequency, perhaps an FM radio station. (The steady frequency is generated inside the radio and is called the "local frequency" and the circuit that generates it is the "local oscillator.") Mixing is more complicated than just adding the signals together. The signals are passed through a device that produces an output that varies in a non-linear way with voltage, like a forward-biased diode. It turns out that several frequencies then appear at the output, one of which is the difference between the signal frequency and the local frequency. (See Horowitz and Hill for details.) Removing the other frequencies with a band-pass filter leaves just this difference frequency. In practice, the local frequency is tuned for each signal frequency so that the difference frequency is always the same. This allows all further processing to be done around a single frequency, so only one filter and one amplifier (called the IF amplifier for "intermediate frequency") are needed to cover the entire FM spectrum.

Experienced hams can create a home-brew Doppler satellite detector using readily available components. Begin by mixing the output of a tunable short-wave receiver and a voltage controlled oscillator in a "balanced" mixer (meaning the mixer outputs only two frequencies, the sum and the difference of the inputs). The output of the mixer must be low-pass filtered to remove the sum frequency. Although Lovberg and Burkhardt's method of measuring the frequency shifts was clever in its time, there are much better ways to do this today. I suggest installing a frequency-to-voltage converter and then passing the output directly to your home computer through an analog-to-digital converter board.

Both Analog Devices (*www.analog.com*) and National Semiconductor (*www.national.com*) manufacture a wide selection of all of these units as complete IC packages. I suggest you check out their Web sites to see what fits your needs, ambition and budget.

One more thing . . . The text does not explain why you need to know the slope of the frequency vs. time curve at its steepest point. Here's the reason. The line tangent to the curve will have the steepest slope at the point at which the frequency is changing the fastest. And the frequency changes the fastest just when the satellite is closest to you. If you've ever listened to the pitch of a train whistle drop as the locomotive passes by you will recall that the pitch drops distinctly from higher to lower just as the train passes because that's the instant at which the train stops moving towards you and starts moving away.

Ed.

22 PREDICTING SATELLITE ORBITS

Conducted by C. L. Stong, May 1974

A mateurs who are interested in predicting the positions in orbit of the earth satellites can now make a device that will perform the calculations automatically. The device is designed for satellites in orbit within 2,000 miles of the earth, and a separate device must be made for each satellite. Details of the apparatus are described by William K. Widger, Jr., who was a pioneer in the development of weather-satellite technology.

"The device consists of a map in the form of a polar projection of the Northern or Southern Hemisphere. A covering sheet of clear plastic is loosely pinned to the center of the map. The pin is fixed at the North or the South Pole.

"The plastic overlay rotates around the pin. The track on the earth's surface over which the satellite passes during any orbit is plotted as a curved line on the overlay [see illustration on page 177]. By rotating the overlay with respect to the polar projection the observer can determine the geographical position of the track, that is, the subpoints on the earth's surface directly under the satellite, at any instant.

"A pattern of concentric circles can be drawn around any point on the earth's surface to represent the local horizon and to serve in determining the series of points directly under a satellite that correspond to the several angles of elevation above the horizon at which the satellite would be seen. When the overlay is rotated to the position at which the track crosses the pattern of concentric circles, the approximate direction and elevation of the satellite are depicted by the points where the track intercepts the circles. The track can be calibrated in terms of the number of minutes that

176

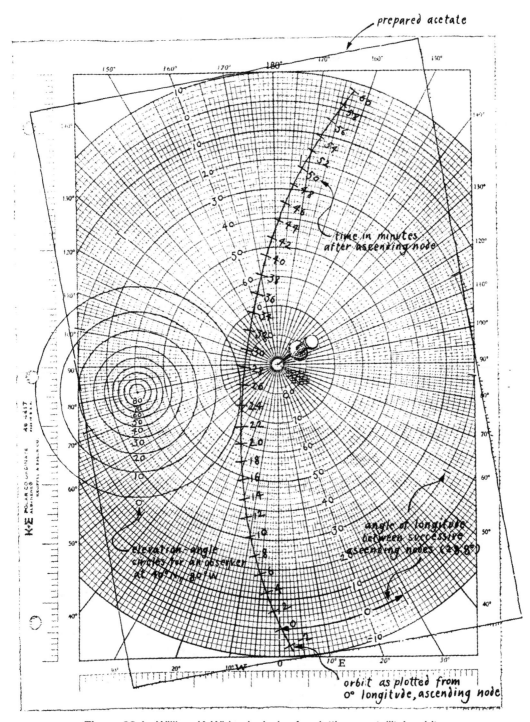

Figure 22.1 William K. Widger's device for plotting a satellite's orbit

have elapsed since the satellite passed over the Equator on its most recent northbound crossing.

"Normally the observer is interested in the position of the satellite in relation to the observing station. For this reason the map need not be cluttered with outlines of the earth's landmasses, although they can be included if the experimenter wants. They are shown on polar projections of the earth in most world atlases.

"For the background map use polar-coordinate graph paper. The track of the satellite is drawn with indelible ink on the type of plastic sheet known as heavyweight clear acetate paper, which is available from dealers in artists' supplies. Mount the graph paper on a stiff backing of cardboard or plywood.

"Begin the project by numbering the major intervals of the graph paper to serve as the geographical coordinates: latitude (ϕ) and longitude (λ). Note that longitude increases both west and east from 0 degrees at the bottom of the map to 180 degrees at the top. Latitude increases from 0 degrees at the Equator to 90 degrees at the Pole. By convention positive numbers designate northern latitudes, negative numbers southern lati-

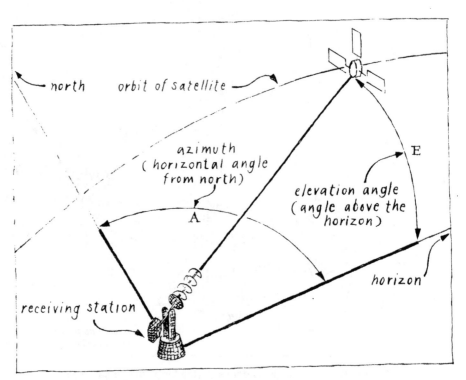

Figure 22.2   Angles related to a position of a satellite

tudes. In making calculations it is convenient to express longitude in terms of a full circle of 360 degrees, beginning at the bottom of the map and proceeding clockwise. Conversion from one system of designating longitude to the other is easy. Subtract longitudes greater than 180 degrees from 360. Label the difference longitude east. For example, longitude 210 degrees is equivalent to 360 − 210 = 150°E.

"Next select and identify (by placing a dot on the graph paper) the location where observations will be made. In the accompanying example [*page 177*] I placed the dot at latitude 40°N and longitude 80°W. I then determined the radii of the concentric circles to be inscribed around this point for indicating the horizon and the elevation angles. To simplify the calculations I assume the earth to be spherical and smooth.

"The accompanying diagram [*bottom of page 178*] depicts the earth as it might appear in cross section if it were cut through the meridians 80°W and 100°E. Note that a straight line tangent to the earth's surface at 40°N defines the horizon and is intercepted by the orbit of the satellite at distances from the observer of 35.6 degrees of arc. The points of interception mark the limits beyond which the receiving station could not pick up line-of-sight signals from a weather satellite that orbits at a height (h) of 1,464 kilometers (910 miles) above the earth's surface. The diagram is based on the U.S. weather satellite NOAA-2, which derives its acronym from the National Oceanic and Atmospheric Administration.

"Observe that an oblique triangle can be drawn by connecting with straight lines the center of the earth, the observing station and the satellite [*see illustration on page 180*]. In this triangle the lengths of two sides are known. One side is the earth's radius, R. In the calculations that follow R is assigned the value of 6,378 kilometers. The other known side is equal to the sum of $R + h$, 7,842 kilometers. The largest angle of the triangle, the one that faces the large known side, is also known. This angle is the sum of the assumed angle of elevation and the right angle made by the vertical line through the station and the tangent to the earth's surface.

"With this information one can easily determine the size of the angle (B) between the known sides and therefore the length of the arc at the earth's surface that is subtended by the known sides. This arc comprises the radius of the circle of points that could lie directly under the satellite when the satellite is at the assumed angle of elevation above the horizon. For example, the triangle that has been drawn with the darkest lines in the diagram corresponds to an elevation angle of 20 degrees.

"The sequence of arithmetical operations required to determine the corresponding radius (B) in degrees of arc for a circle of subpoints at 20-degree elevation is specified by the first of the accompanying formulas [*page 181*]. Divide the earth's radius by the sum of the radius and the

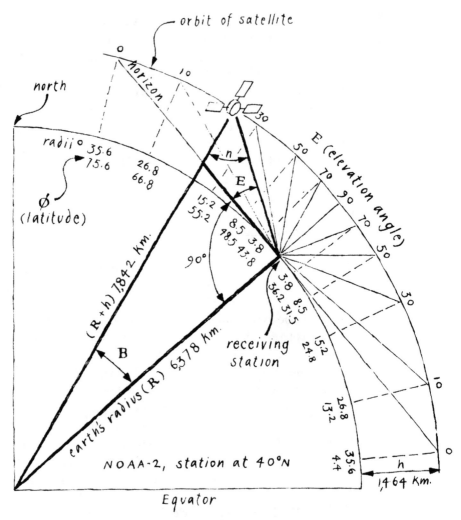

Figure 22.3 Geometry of elevation circles

height of the satellite: 6,378 / (6,378 + 1,464) = .8133. From a table of trigonometric functions determine the cosine of each desired angle of elevation. In this example the desired elevation is 20 degrees. According to the table, cos 20° = .9397. The formula requires the product of $[R/(R + h)]$ × cos E, which in this example is .8133 × .9397 = .7642.

"Next the formula requires the angle (arc cosine) that corresponds to this cosine. According to the trigonometric table, the cosine .7642 corresponds to the angle 40.2 degrees. [Alternatively, you can dispense with the table if you have a calculator or spreadsheet that supports the 'acos' func-

$B = \text{arc cos } [\{R/(R + h)\} \cos E] - E$	(1)
$i = \text{arc cos } [-4.7349 \times 10^{-15} (R + h)^{7/2}]$	(2)
$I = 180° - i$	(3)
$P = 1.6586 \times 10^{-4} (R + h)^{3/2}$	(4)
$\theta = \text{arc sin } (\sin I \sin L)$	(5)
$\lambda = \{\text{arc tan } (\cos I \tan L)\} + PL/1{,}440$	(6)
$\theta_t = \text{arc sin } [\sin I \sin (360t/P)]$	(7)

Table 22.1 Formulas for predicting an orbit

tion. Ed.] Finally, the formula states that the desired radius (B) in degrees of arc is found by subtracting the angle E (20 degrees) from the arc cosine angle: $B = 40.2 - 20 = 20.2°$.

"To draw on the map a circle that represents points in all directions that are 20 degrees above the local horizon, place the point of a pen compass on the equator of the graph paper at 0 degrees longitude and open the instrument until the nib of the pen corresponds to latitude 20.2 degrees. With this distance as the radius, transfer the point of the compass to the location of the observing station on the map and draw the circle. Similarly determine the radii and inscribe circles at other angles of elevation as desired. In the case of NOAA-2 I have tabulated radii for elevation angles at intervals of 10 degrees from 0 to 90 degrees [*see table on page 182*].

"Place the clear plastic overlay over the map and pin it at the Pole so that the overlay can be rotated around the pin. The path of the satellite will be plotted on the overlay in terms of the object's changing latitude and longitude. These coordinates can be calculated for a track of any length in increments of, say, 10 arc degrees, beginning on the Equator at 0 degrees longitude. Each length will be the hypotenuse of a right triangle as drawn on the surface of a sphere. The lengths of the corresponding legs of the triangle are measured in degrees of latitude and longitude.

"The fact that members of the solar system rotate in step somewhat like the meshed gears of a machine enables one to plot the course of an earth satellite with a surprisingly small amount of initial information. For example, all that is needed to plot the course of NOAA-2 is the fact that it travels 1,464 kilometers above the earth's surface in a sun-synchronous orbit. The plane of a sun-synchronous orbit turns one full revolution with respect to the sun in a year of 365.2422 days. The angle made between the orbital plane and the sun remains constant. Moreover, the ascending node (the point at which a northward-bound earth satellite crosses the plane of

ELEVATION CIRCLES		SATELLITE TRACK DATA						TIME FROM ASCENDING NODE					
E	B	L	θ	λ	L	θ	λ	t	θ_t	t	θ_t	t	θ_t
0	35.6	-20	-19.6	-5.8	100	74.7	139.1	-6	-18.3	18	54.5	42	47.4
10	26.8	-10	-9.8	-2.8	110	67.0	159.7	-4	-12.2	20	60.3	44	41.4
20	20.2	0	0	0	120	58.0	170.3	-2	-6.1	22	65.9	46	35.4
30	15.2	10	9.8	2.8	130	48.6	176.8	0	0	24	71.1	48	29.3
40	11.5	20	19.6	5.8	140	39.0	181.6	2	6.1	26	75.5	50	23.2
50	8.5	30	29.3	9.1	150	29.3	185.3	4	12.2	28	78.1	52	17.1
60	6.0	40	39.0	12.8	160	19.6	188.6	6	18.3	30	77.7	54	11.0
70	3.8	50	48.6	17.6	170	9.8	191.6	8	24.4	32	74.7	56	4.9
80	1.9	60	58.0	24.1	180	0	194.4	10	30.5	34	70.1	58	-1.2
90	0	70	67.0	34.7	190	-9.8	197.2	12	36.6	36	64.8	60	-7.4
		80	74.7	55.3	200	-19.6	200.2	14	42.6	38	59.2	62	-13.5
		90	78.3	97.2	210	-29.3	203.3	16	48.6	40	53.3	64	-19.6

Table 22.2 Data for a plotting device designed for the satellite NOAA-2

the celestial equator—not, as in the usual case, the plane of the ecliptic) occurs at the same local time on each orbit.

"Offhand one might suppose that the orbit of any earth satellite would remain stationary in relation to the fixed stars. It probably would if the universe were symmetrical. Actually the orbits of earth satellites are strongly influenced by the aspherical shape of the earth, which bulges slightly at the Equator. Satellites that orbit in planes making an angle other than 0 or 90 degrees with respect to the earth's Equator respond to the gravitational pull of the bulge much as a gyroscope behaves when it is supported at one end of its shaft. Instead of falling the gyroscope rotates around the point of its support. In the same way the gravitational force between the bulge of the earth and the satellite causes the orbit of the satellite to precess in relation to the fixed stars. This effect is observed in the wobbling motion of an inclined top. Indeed, the rate at which the orbital plane of an earth satellite rotates varies with both the altitude of orbit and the angle at which the plane of the orbit is inclined with respect to the Equator.

"To establish the sun-synchronous condition, weather satellites such as NOAA-2 are injected into orbit at an extremely precise angle of inclination, which is based on mathematical relations worked out by specialists in astrodynamics. If the height of the orbit is known, the sun-synchronous angle of inclination (i) can be calculated, as shown in Formulas 2 and 3 [*table on page 181*]. In the case of NOAA-2 the inclination of the orbit (i) is equal to the angle (arc cosine) that corresponds to the cosine of $-4.7349 \times 10^{-15} \times (R + h)^{7/2}$. To find the 7/2 power of $R + h$ for NOAA-2, multiply by itself seven times the square root of the sum of $6,378 + 1,464$. [Again, a modern calculator makes this easy if it supports the X^Y function. Ed.] The square root of the sum is 88.56, and 88.56^7 is 4.2723×10^{18}. The product of $4.2723 \times 10^{13} \times -4.7349 \times 10^{-15}$ is $-.2022$.

"The angle that is equivalent to the negative cosine $-.2022$ is found by subtracting from 180 degrees the angle that corresponds to (positive) cosine .2022: arc cos $.2022 = 78.33$; arc cos $-.2022 = 180° - 78.33° = 101.67°$ $= i$, which is the desired angle of inclination of the satellite NOAA-2. Formulas that determine the track (L) on the surface of the earth over which satellites such as NOAA-2 orbit make use of the supplement (l) of the angle of inclination: $180° - 101.67° = l$.

"Also required to calculate the track (L), as we have seen, is the period of time in minutes (P) necessary for the satellite to complete one orbit of the earth beginning and ending at the ascending node. The period (P), as indicated by Formula 4, is equal to 1.6586×10^{-4} multiplied by the cube of the square root of $R + h$. For NOAA-2 the square root of $R + h$ is 88.56. The cube of 88.56 is 6.9456×10^5. Therefore the nodal period of NOAA-2 (P) is $1.6586 \times 10^{-4} \times 6.9456 \times 10^5 = 115.19$ minutes.

"With this information one can calculate intervals of latitude (ϕ) and longitude (λ) that comprise the legs of the spherical triangle that has as its hypotenuse the track (L). When I design a prediction device, I normally begin by assuming that the track starts on the Equator at 0 degrees longitude. I then compute the corresponding latitude and longitude as the satellite advances along its track in increments of 10 degrees of arc. A somewhat more accurate curve could be plotted by reducing the increments to five degrees of arc, which would double the number of computations. [Think spreadsheet! Ed.]

"Latitude (ϕ) is determined with the aid of Formula 5. For example, when NOAA-2 has orbited through a distance of 10 degrees of arc from its assumed starting point above the Equator, it arrives at the latitude: arc sin (sin 78.33° × sin 10°) = .9793 × .1736 = .17 = 9.8° = ϕ. After NOAA-2 has advanced 90 arc degrees along its track its latitude has increased to arc sin (sin 78.33° × sin 90°) = .9793 × 1 = .9793 = 78.33°, its closest approach in latitude to the North Pole. When NOAA-2 completes half of its orbit (180 degrees), its latitude falls to arc sin (sin 78.33° × sin 180°) = .9793 × 0 = 0°. Stated differently, the satellite has arrived above the Equator on the opposite side of the earth. From 180 to 360 degrees the trigonometric sign of all angles is negative, indicating that these latitudes lie in the Southern Hemisphere.

"Each increment of longitude (λ) constitutes the second leg of the associated spherical triangle. The increments are calculated with the aid of Formula 6. Longitude is generated by two motions: the westward advance of the satellite on its retrograde orbit and the slower motion of the observer as the earth's rotation carries the station eastward. The component of the longitudinal motion for which the satellite is responsible is described by the formula as an arc that corresponds in degrees (arc tangent) to a trigonometric tangent expressed as the product of cosine I multiplied by tangent L. To this arc, according to the formula, must be added the second component of motion: an arc equal in degrees to the product of $P \times L/1,440$.

"Assume that NOAA-2 has advanced 10 degrees along the orbit (along the hypotenuse of the spherical triangle). It has reached the latitude 9.8°N, as has been previously calculated. Simultaneously the satellite has moved westward along the inclined orbit: $L = 10°$. As has been previously determined, the cosine of I (78.33 degrees) is .2022. The trigonometric table lists the tangent of 10 degrees as .1763. Substitute these values in the formula: arc tan (.2022 × .1763) = arc tan .0356 = 2°. This is the arc through which NOAA-2 would have orbited in longitude if the earth were not rotating.

"The motion of the earth generated an interval of longitude: $P \times L/1,440 = 115.19 \times 10°/1,440 = .8°$. The total increase in longitude is therefore $\lambda = 2 + .8° = 2.8°$. Near the Equator, where the meridians are widely spaced, the rate at which the satellite moves westward is low in relation to its motion northward. Over the first 10 degrees of NOAA-2's orbit from the Equator its longitude increased by only 2.8 degrees, an average of only .28 degree of longitude per degree along its track.

"This rate increases dramatically as the satellite approaches the poles, where the meridians meet. When L is 90 degrees, the tangent of L is infinity. The product of cosine I times infinity is infinity. The arc tangent of infinity is 90 degrees. At this point along the track the component of longitude generated by the earth's rotation is 115.19° × 90/1,440 = 7.2°. Therefore $\lambda = 90° + 7.2° = 97.2°$. When L has increased to 100 degrees, the equation is tan $L = 180° - 100° = $ tan 80°. According to the laws of trigonometry, tangents of angles that lie between 90 and 180 degrees are negative numbers. Hence tan 100° = –tan 80° = –5.6713. Substitute the appropriate values in the formula, cosine I and tangent L: .2022 × –5.6713 = –1.1467. Therefore at this point along the track the component of arc for which the satellite is responsible is 180° + arc tan –1.1467 = 180° – 48.9° = 131.1°.

"To this longitude must be added the increment contributed by the earth's rotation: 115.19 × 100°/1,440 = 8°. Hence after NOAA-2 has orbited 100 degrees from the Equator its longitude has increased to $\lambda = 131.1° +

8° = 139.1°. During the 10 degrees of arc along the track from 90 to 100 degrees the longitude of the satellite shifted westward from 97.2 to 139.1 degrees, an advance of 41.9 degrees compared with the difference of only 2.8 degrees for the same length of track at the Equator!

"When you do the arithmetic, keep in mind that the tangents of angles that lie between 180 and 270 degrees are positive. The calculation of these angles is similar to that of angles between 0 and 90 degrees. The accompanying table lists the latitudes (ϕ) and longitudes (λ) that determine points at each 10 degrees of arc along the track of NOAA-2. Beginning on the Equator, plot the points on the plastic overlay. (The overlay can be anchored temporarily at the edge of the map with bits of adhesive tape.) Connect the points by a graph to represent the track. The graph can be drawn smoothly in indelible ink with a ruling pen and a drafter's French curve.

"The track can be calibrated in minutes of time since the satellite crossed the ascending node. Intervals of two minutes are convenient. I prefer to plot the intervals in terms of latitude (ϕ_t) with the aid of Formula 7. For example, two minutes after NOAA-2 crosses the Equator its latitude, according to the formula, is arc sin [sin 78.33 × sin (360 × 2/115.19)] = arc sin .1066 = 6.1° = ϕ_t. If P is known in minutes, the formula can be simplified. Divide 360 by the known period. Thereafter multiply the intervals of time (t) by the resulting constant. For NOAA-2 the constant is 360/115.19 = 3.125. The formula becomes arc sin (sin I sin 3.125 t). The accompanying table [*top of page 182*] lists latitudes for calibrating the track of NOAA-2 during the first hour of its orbit from the ascending node.

"If two or more tick marks are inscribed on the overlay at the Equator, the device will serve for predicting the location of the satellite for as many as about 12½ orbits. The marks are spaced at intervals of longitude equal to the satellite's apparent westward movement during each full orbit, as measured from consecutive ascending nodes. This distance, in degrees of arc, is about .25 times P. For NOAA-2 the tick marks are spaced at intervals of .25 × 115.19 = 28.8°. In general predictions of more than 24 hours are not useful because small errors build up rapidly.

"To use the instrument rotate the overlay to the point at which the zero time of the track coincides with the longitude of the satellite at its next ascending node near the longitude of the observer. This information is broadcast worldwide by radio station W1AW, as Eugene F. Ruperto explains in his description of an amateur weather-satellite station in this department [*Scientific American*, January 1974].

"The fact that the orbit of the satellite can penetrate the observer's horizon as plotted on the map, or even some of the inner circles of elevation, does not necessarily ensure that radio signals from the satellite can be picked up. The device is designed on the unrealistic assumption that the earth is a smooth sphere. Local features of the terrain may interfere with the line-of-sight path. Experience at a given location will soon determine for the observer the bands of ascending-node longitudes and the corresponding interval of time after the ascending node when the satellite is within receiving range of the station."

PART 4

THE PLANETS,
COMETS,
AND
STARS

23 AMATEUR OBSERVATIONS OF JUPITER

Conducted by Albert G. Ingalls, May 1953

Unlike Mars the planet Jupiter has never been seen. What we see of Jupiter is not the planet itself but an unbroken canopy of banded clouds hiding some wholly unknown entity beneath. The spectrograph tells us that these clouds consist of ammonia and probably methane, and the thermocouple reports that their temperature is more than 200 degrees below zero Fahrenheit.

A six-inch telescope magnifies this cloud-enveloped body to the apparent size of our moon as seen with the naked eye. A twelve-inch telescope can give a clearer image. The two most prominent dark bands, a number of fainter dark bands and certain markings near the bottom and around the polar regions are red-brown clouds. The bright bands in between are white or yellowish-green clouds. The eye-shaped object called the Great Red Spot, which has been observed closely since 1878, can also be seen. Sometimes a dark spot can be seen above it. This is the shadow of Jupiter's satellite Ganymede.

None of the visible features of Jupiter is fixed. The cloud bands vary from year to year—in number, in width and in latitude. Most of them last only a few weeks or months. Each rotates at a different rate. The broad bright equatorial band rotates most rapidly: its period is 9 hours and 50 minutes. The periods of the others are five or six minutes shorter; each rotates at a different rate without system or relation to latitude. Their edges are sharply bounded. Each drifts slowly past its neighbor.

Within the bands there are many minor markings, and these continually change. Watching the changes provides such variety that the observation of Jupiter is a lively business, especially since the planet rotates so rapidly. However, the observation would soon lose interest were it not for the intriguing riddle beneath, to which the puzzling visible performances seem in some mysterious way related. From the known mass and volume of Jupiter it is easy to calculate that the density of the planet averages but one and one third times that of water, and from this and the gradual shifting of the clouds it has been conjectured that the planet is partly fluid.

Almost the only systematic observers of Jupiter's ever-changing clouds have been serious amateurs, working mainly in organizations. Such work was begun several decades ago in Great Britain and has been taken up also in the U.S. and Canada. The observers have accumulated detailed records of the dark belts, the intervening bright zones and the many markings within each. These have been published as occasional reports in the memoirs of the British Astronomical Association. In the U.S., observers' reports and drawings of changes are collected by the Jupiter Recorder for publication in *The Strolling Astronomer,* the periodical of the Association of Lunar and Planetary Observers, which any amateur may join [*www.lpl.arizona.edu/alpo/*].

One ALPO member who has done outstanding work in observing Jupiter and in using observational data accumulated by others is Elmer J. Reese of Uniontown, Pa. Reese selected two of Jupiter's bands in which changes recorded since 1940 had persisted uncommonly long, and minutely observed them himself between 1940 and 1951 with his homemade six-inch reflector. A brief analysis of his findings has been published in *The Strolling Astronomer,* and a longer article on them has been written by Walter H. Haas, then ALPO director and editor, for *The Griffith Observer,* publication of the Griffith Observatory in Los Angeles.

Reese's chosen bands are the ones between the shadow of Ganymede and the Great Red Spot. In the standard nomenclature these are respectively the South Temperate Zone and the South Temperate Belt (south because astronomical telescopes invert). Like all other zones and belts on Jupiter, these encircle the planet.

The drawing on page 191 shows the bands unrolled. The dusky lower section in each band is the South Temperate Belt. The brighter upper part is the South Temperate Zone. The elongated white markings BC, DE, and FA show eruptive disturbances from beneath. They have persisted for several years longer than any disturbance recorded on Jupiter except the Red Spot and a nearby eruption which lasted for several decades.

Reese watched these white sections diminish in length as the dusky

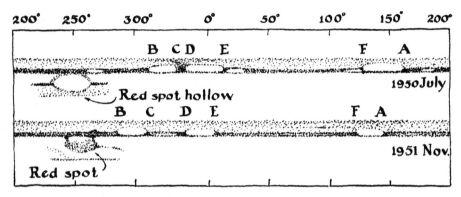

Figure 23.1 Changes during an eruption on Jupiter

sections CD, EF and AB gradually encroached on them during the period indicated. He also saw them drift to the left at differing distances from the Red Spot. This spot lies in a third band called the South Tropical Zone. As the lower drawing shows, the Red Spot is in a depression called the Red Spot Hollow. At times the spot itself disappears.

Reese next plotted drift curves, shown on the next page, for six longitudinal sections—three dusky, three bright. The slope of these curves indicates duration and shows the rate at which the feature is moving. To make measurements he needed a way to time the markings as the planet rotated. While this can be done with a filar micrometer or by measuring photographs, the amateur uses a primitive method which may be equally precise. As a feature approaches the central meridian of rapidly rotating Jupiter, one decides to the nearest minute when it is centered on the disk and records the time. It is no more difficult to estimate accurately when a marking is centered on a planetary disk than to center a picture accurately in a frame with the eye alone. When the same marking is observed for a number of days, any error is reduced proportionately; in a month the error is reduced from minutes to seconds. On the drift curves drawn by Reese the shaded strips represent the three longitudinal dusky sections of the South Temperate Belt and the white strips the bright sections. (The strips should not be confused with the actual bands.) It is easy to see that in 1940 the white sections were wider, longitudinally, than the dusky sections. By 1950 the dusky sections had encroached so far on the bright that the bright eruption had almost ceased to exist.

In 1948 the motion of the six sections, which had been uniform, suddenly decelerated, as shown by the knee in the curve. At that date their rotation period lengthened by four seconds to 9 hours, 55 minutes, 10 sec-

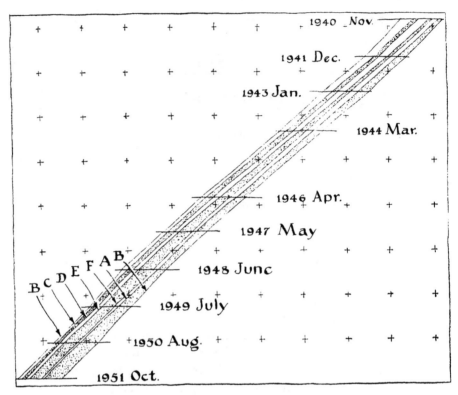

Figure 23.2   Drift curves for a belt on Jupiter

onds. Some unknown influence had also applied a brake to the entire band from the beginning in 1940: in that 11-year period it fell back nine laps (note nine spaces) in the race with adjacent bands.

It is not nearly as easy to keep track of a protean marking on Jupiter at the telescope eyepiece as it is to examine crisp drawings in an armchair. Another source of perplexity is the annual conjunction of Jupiter with the Sun, which makes it invisible for several months. When it comes into view again, the marking under observation may have changed so much that it must be identified by projecting its drift line on the chart. Nor is the seeing always good. Haas pointed out that a telescope of the very best optical quality and 10 or 12 inches in diameter gives the best results in the search for long-enduring markings on Jupiter. He asks, however, whether the scarcity on Jupiter of long-lasting features, such as the eruptions observed by Reese, is real or only apparent: "The ability of the telescopist to fail to see what he is not looking for is at times most remarkable!"

Haas calls Reese's work "an outstanding piece of Jupiter research done by an outstanding amateur astronomer." What have his 11 years of observations proved? Are the eager amateurs who sit up all night and hastily record 100 transits of markings on Jupiter accomplishing anything or merely accumulating useless statistical data on some clouds of gas? The basic data of science have often looked useless until the key to a riddle has turned up. Then the statistical data suddenly become valuable.

24 PREDICTING PLANETARY ALIGNMENTS

Conducted by C. L. Stong, August 1975

From time to time a planet, as it is observed from the earth, moves into alignment with the sun or with another planet. These alignments are termed conjunctions or oppositions, according to the relative positions of the earth and the sun. They can be predicted by a graphical technique that has been devised by B. E. Johnson of Mercer Island, Wash.

Conjunctions and oppositions have attracted interest throughout recorded history. To astrologers they were portents, usually dire. Today the forces that arise from such groupings are taken into account by digital computers to help determine the course of space probes.

To amateur observers conjunctions and oppositions form interesting patterns in the night sky. Moreover, a conjunction marks the onset of spectacular motions that appear to carry the planets involved along a looping path against the background of the fixed stars. The looping motion can be followed by making nightly observations for a few weeks.

No graphical technique can yield an exact prediction. Moreover, planetary motions are not exactly proportional to the passage of time. Even so, Johnson's method generates results of sufficient accuracy to satisfy most amateur needs. Johnson, who is an electrical engineer, describes his scheme as follows.

"Planetary alignments are of four kinds, depending on the distance of the planets from the sun. The two 'inferior' planets, Mercury and Venus, are closer to the sun than the earth is. Mercury completes one orbit around the sun in 87.969 earth-days; Venus, in 224.7 earth-days. These intervals are the sidereal periods of the planets. (The sidereal period of the

Anyone with a computer and a good astronomy software package can generate a schedule of planetary alignments in seconds. However, doing so doesn't help the amateur gain a good intuitive feel for the relative motion of the planets. Johnson's simple graphical method, although not as precise as a high-tech calculation, cuts right to the core physics of planetary motions. In this regard these techniques are quite instructive. I gained a lot by going through this exercise, and I'm sure that many amateur astronomers would also benefit.

Ed.

earth is 365.2563 days. The interval between two successive vernal equinoxes of the earth is the tropical year. It spans 365.2422 days.)

"The sidereal period of each planet increases with the planet's distance from the sun. The inclinations of the orbital planes of the planets differ from one another by only a few degrees, with the exception of the orbital plane of Pluto. The orbit of Pluto is inclined about 17 degrees to the ecliptic, the plane in which the earth travels.

"An apparent alignment of the inferior planets occurs when Mercury or Venus moves exactly between the earth and the sun. (Sometimes both of them move.) These events are called inferior conjunctions. Continued rotation carries each planet at its characteristic velocity to the opposite side of its orbit to points where the sun lies in the line between the planets and the earth. These positions are termed superior conjunctions.

"The 'superior' planets (Mars, Jupiter, Saturn, Uranus, Neptune and Pluto) are, in that order, increasingly distant from the sun and have correspondingly longer sidereal periods. A superior planet reaches conjunction when it moves to the point where the sun lies between the planet and the observer's meridian at noon. At that position the planet is lost from view in the sun's glare.

"Continued rotation carries each superior planet to the point where the earth is eventually aligned between the planet and the sun. The planet is then said to be in opposition. The position appears on the observer's meridian at midnight. Periodically several superior planets reach opposition simultaneously. Inferior planets can never reach opposition, and superior planets cannot reach inferior conjunction [see *the illustration on page 196*].

"The interval of time between successive conjunctions or successive oppositions of a planet is its synodic period. This period can be calculated easily by either of two methods. For a superior planet the reciprocal of the synodic period is equal to the difference between the reciprocal of the

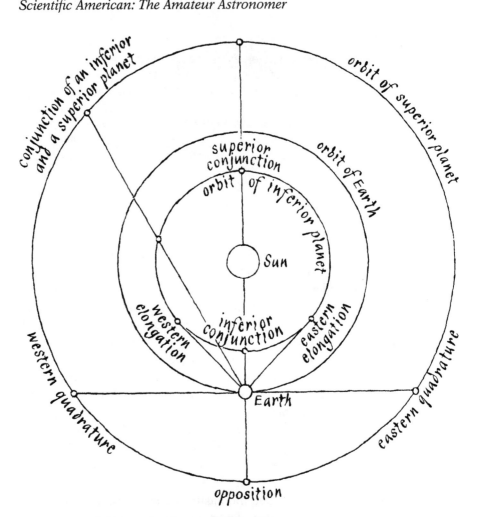

Figure 24.1 Orbits, conjunctions and oppositions

earth's sidereal period and the reciprocal of the planet's sidereal period. With Mars as an example the reciprocal of the synodic period is $1/365.25 - 1/686.98 = 1/779.9$. The interval between successive conjunctions or successive oppositions of Mars is 779.9 earth-days.

"The synodic period of the inferior planets is given by the difference between the reciprocal of the planet's sidereal period and the period of the earth. For Venus the reciprocal of the synodic period is $1/224.7 - 1/365.25 = 1/583.94$. The interval between successive superior or inferior conjunctions of Venus is 583.94 earth-days.

"A simple graph can also display the synodic period of a planet [*see the illustration on page 197*]. Draw a pair of rectangular coordinates. Calibrate the abscissa in intervals of one year for, say, 2½ years and the ordinate in units of 360 degrees for at least 720 degrees. The fact that the earth completes an orbit around the sun (traverses 360 degrees) in one year can be represented by a graph in the form of a diagonal line drawn from zero on the abscissa to intercept 360 degrees at one year. The slope of the graph is equal to the quotient of 360 degrees divided by the number of days in a year, or .9856 degree per day.

"The graph of any planet can be similarly plotted on the same coordinates. For example, the motion of Mars can be represented by a straight line that begins at zero on the abscissa and intercepts the line representing 360 degrees at the sidereal period of Mars (686.99 earth-days). The slope of the Martian graph is equal to the quotient of 360 degrees divided by the number of earth-days in one sidereal year of Mars, or .5240 degree per day. The difference in the rate at which the two planets travel around the sun is indicated by the difference in the slopes of the graphs.

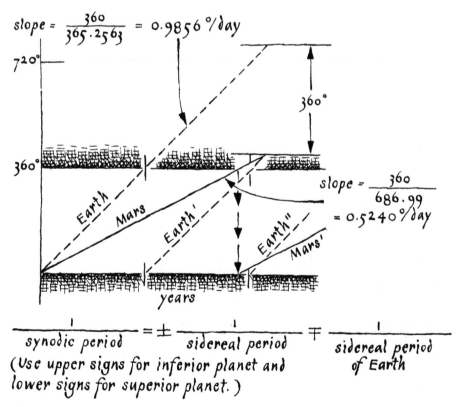

Figure 24.2 Basic graph of planets in orbit

"Periodically the difference between the graphs as measured on the ordinate amounts to a multiple of 360 degrees. The multiples occur at time intervals equal to the synodic period of the planet and are equal to 360 degrees divided by the difference between the slopes of the graphs: $360/(.9856 - .5240) = 779.9$, which is the synodic period of Mars. Graphically the synodic period can be measured along the abscissa. It amounts to 2.135 sidereal earth-years.

"A simple modification can adapt this graphical technique for displaying both past and future conjunctions and oppositions. Limit the ordinate to one interval of 360 degrees. Divide the abscissa into any number of yearly intervals. The accuracy of the predictions increases with the size of the graph.

"The graph of each planet originates on the abscissa at the beginning of the sidereal period and terminates at the point where the end of the period intercepts 360 degrees as measured by the ordinate. Continued motion of the planet is depicted by initiating a new graph at zero degrees and at the first day of the next sidereal interval as suggested by the broken line 'Earth" and the solid line 'Mars" in the illustration. The result is a saw-tooth pattern on the graph of each planet.

"The slope of the 'teeth' and their number vary with the sidereal period of each planet. The accompanying graph [*below*] depicts motions of the

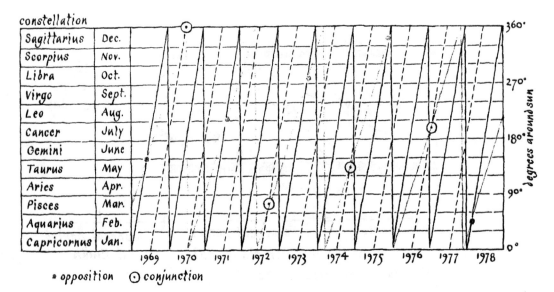

Figure 24.3 B. E. Johnson's graphical way of displaying conjunctions and oppositions

earth and Mars through the decade ending in 1979. The ordinate of the graph is divided into 12 intervals of 30 degrees. Each interval represents the partial orbit of the planet around the sun in roughly one month. Each interval of 30 degrees is labeled to indicate approximately the constellation in which the sun appears at the time and also the corresponding months during which the constellations appear on the meridian.

"The abscissa of the graph is divided into 10 intervals that represent the years 1969 through 1978. To avoid cluttering the graph with fine detail the yearly intervals have not been subdivided into intervals representing months. A black line that represents the motion of the earth for 1969 is drawn on the graph beginning on January 1, 1969, at zero degrees and sloping diagonally to December 31 at 360 degrees.

"A line begins at approximately 75 degrees on the ordinate and January 1 on the time scale. This graph depicts the motion of Mars. It intercepts the time scale approximately 548 days later (about July 2, 1970). On that date the new graph of Mars begins at zero degrees, as measured along the ordinate. In contrast, graphs of the earth's motion begin and end with each calendar year.

"Note the periodic intersections of the two lines every 779.9 earth-days. The intersections mark the dates when Mars is in opposition, as it will be for example on December 15, 1975. Conjunctions can be depicted by displacing the graph of the earth's motion by an interval of six months with respect to the abscissa, as indicated by the broken line. Dates when Mars has been in conjunction are indicated by the intersections of the broken and the solid lines.

"As a convenience I rule a pattern of horizontal and vertical lines 9¾ by 4½ inches in size on standard 11-by-8½-inch paper and label the 12 horizontal lines with the months in the second column from the left [*see the illustration on page 200*]. At the right margin the same 12 lines are labeled in multiples of 90 degrees from zero through 360 degrees. Duplicates of this form are made on an office copying machine. The accompanying drawing of the form depicts the graphs of the nine planets. It illustrates how the display tends to become crowded with planets within 200 million miles of the sun as well as those that are on the order of a billion or more miles away.

"The slope of any graph of this kind can be plotted most accurately by locating the origin at zero degrees and zero years and terminating it either at the intersection of the planet's sidereal period and 360 degrees or at the point on the orbit (expressed in degrees) that the planet would reach within the maximum number of years displayed by the abscissa. In the accompanying example the slopes of the graphs of Mercury, Venus,

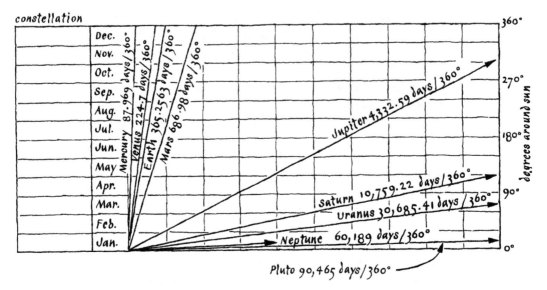

Figure 24.4   Relative slopes of all planets

the earth and Mars were determined by drawing straight lines between the point of origin and the 360-degree points that corresponded to the respective sidereal periods of the four planets within 200 million miles of the sun.

"In this example the abscissa spans 10 years, or 3,652.5 days. To what point along the right margin of the chart should a graph be drawn to represent the slope of Pluto? The sidereal period of Pluto is 90,465 earth-days. In one day Pluto would rotate 360/90,465 = .0039794 degree. In 10 years it would orbit through 3,652.5 days multiplied by .0039794 degree per day, or 14.53 degrees. A straight line drawn from this point on the right margin of the coordinates to the point of origin at the left displays the slope of Pluto's graph.

"Because the synodic period is a function of the sidereal period, I calibrate the coordinates of the graphs in terms of the sidereal year of 365.2563 days and initiate the orbit at zero degrees on January 1 instead of at the vernal equinox, as would be done in the case of the tropical year of 365.2422 days. With the slopes determined, the graphs of the planets must then be displaced in time from the point of origin as required to locate the intersections of the planet and earth graphs at the exact dates of known conjunctions and oppositions. Almanacs list the dates of these events for each calendar year. A reference that is remarkably complete and easy to use is *The Astronomical Almanac*. The volume is updated and

PLANET	ANGULAR VELOCITY IN ORBIT (DEGREES PER EARTH-DAY)	SIDEREAL PERIOD (EARTH-DAYS)	SYNODIC PERIOD (EARTH-DAYS)
MERCURY	4.092339	87.96	115.88
VENUS	1.6021	224.70	583.92
EARTH	.985609	365.25	—
MARS	.524033	686.98	779.94
JUPITER	.083091	4,332.59	398.88
SATURN	.033460	10,759.22	378.09
URANUS	.011732	30,685.40	369.66
NEPTUNE	.005981	60,189.00	367.49
PLUTO	.003979	90,465.00	368.73

Table 24.1 Table of planetary motions

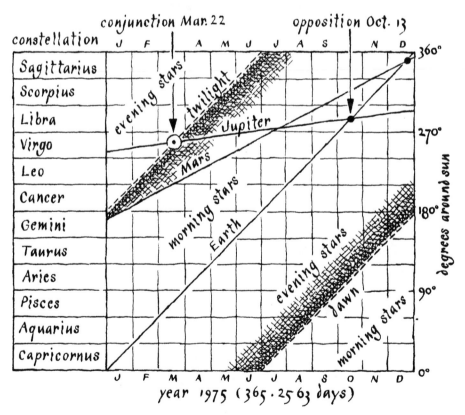

Figure 24.5 Prediction of morning and evening events

201

reissued annually. It can be purchased from the Superintendent of Documents (U.S. Government Printing Office, Washington, D.C. 20402).

"Having plotted graphs of the earth's motion for as many yearly intervals as are provided by the coordinates, I put one or more points on the graphs to indicate oppositions of the planet to be plotted. The graph of the planet is drawn through these points at the slope previously determined for the planet. A line that has been drawn lightly from the zero point to determine the slope of a planet can be conveniently transferred to intercept the graph of the earth at points of known conjunctions or oppositions by the instrument known to navigators as the parallel ruler.

"Planets can be in conjunction with each other as well as with the sun. These events appear as intersections between the graphs of the planets at points removed from the graph of the earth. For example, the accompanying graph [*page 201*] indicates that Mars and Jupiter were in conjunction about July 2, 1975. Because their intersection appears above the graph of the earth they were seen as morning stars. Evening events appear below the graph of the earth's position."

25 CATCH A COMET BY ITS TAIL

By Shawn Carlson, January 1997

On March 9, 1997 beginning 41 minutes after midnight Universal Time, a few hardy souls who were willing to brave the Siberian winter witnessed a total eclipse of the sun. As the lunar shadow rushed northward across the subzero landscape, intrepid observers saw, in addition to the usual spectacular solar corona, a streak of light painting the darkened sky. Comet Hale-Bopp (known to astronomers as C/1995 O1), the brightest comet in more than two decades, was just 22 days away from perihelion and only 13 days short of its closest approach to the earth. Its brilliantly illuminated tails produced a dazzling display.

If a trek to subarctic Siberia didn't fit your plans, Hale-Bopp still provided you with sensational views no matter where you lived. It also offered amateurs a chance to contribute to cometary research: the Harvard-Smithsonian Center for Astrophysics in Cambridge, Mass., coordinated a global net of observers, and many amateurs participated.

Astronomers were agog over Hale-Bopp because its nucleus was particularly active. A cometary nucleus is a fluffy ball of ice and rock whose surface evaporates as the wanderer nears the sun. The resulting streams of dust and gas make up the comet's tails. (The glowing dust traces a curved path; ionized gas travels in a straight line away from the sun.) Hale-Bopp's nucleus first began spurting out visible jets of debris as it passed the orbit of Jupiter, roughly seven astronomical units from the sun (1 AU is the average distance from the sun to the earth, or about 150 million kilometers). Experienced naked-eye observers started watching Hale-Bopp in

May 1996 (most comets are visible to the unaided eye only a few months before perihelion), and it became visible to the rest of us in January 1997.

Comet Hale-Bopp is now long gone. But one never knows when another bright comet will grace our skies. Indeed, a new comet that is visible with the naked eye makes an unexpected appearance every few years, and amateurs can always make important observations that can contribute to our understanding about these celestial vagabonds. So here are some basic techniques you can use to get involved the next time a comet wanders by.

A comet's tails (one dust, the other gas) reveal some of its most intimate secrets of composition and structure. They also give earthbound watchers a fine traveling laboratory to chart the solar wind. Tails are sometimes decorated with feathery features that flow outward under the solar wind's influence. Comet Kohoutek delighted astronomers in 1974 with at least two prominent examples of these skirting disturbances.

Any amateur can record these and other features of the comet's tails. First, you'll need a good star atlas that maps stars in terms of right ascension and declination. (*Norton's 2000.0 Star Atlas and Reference Handbook*, 18th edition, by Ian Ridpath, is probably best for this purpose.) You'll also need a drafting compass and a large bow-shaped angular scale. The bow, made from a flexible meter stick or yardstick and a long piece of scrap wood, will let you locate the tails' features to about 0.1 degree of arc.

Sketch the tails directly on the appropriate page of the star atlas (or on a good photocopy). Locate the comet's head by measuring the angular separation between the head and the three nearest stars in the atlas. Celestial maps mark the positions of stars in terms of declination and right ascension; to convert from angles to distance on the page, note that one hour of right ascension equals 12 degrees at the celestial equator. Elsewhere in the sky, divide the distance at the equator by the cosine of the declination.

For each measurement, set the compass to the appropriate opening and scribe a small arc through where you expect the head to be. The precise location of the head is where the arcs intersect. Follow the same procedure to mark all the other major features in the tails and then fill in the finer details. Use a telescope or binoculars for this part. By carrying out this procedure every clear night, you can document the evolution of the tails.

The comet's fuzzy head, or coma, also changes over time in size, brightness and degree of condensation. The best way to measure its size is to use an eyepiece with a calibrated scale etched into the lens. They're a bit pricey, but for accuracy they can't be beat. Check out the micro guide eyepiece from Celestron in Torrance, Calif. (*www.celestron.com*); call (800)

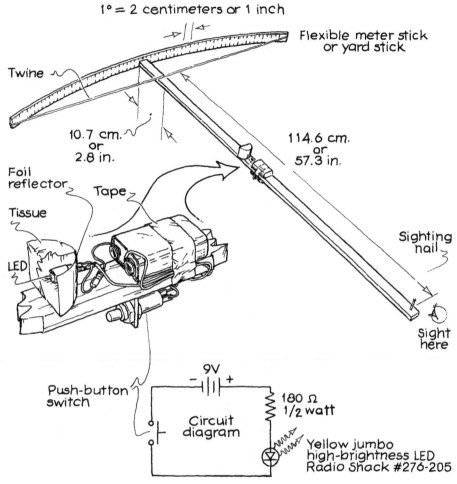

Figure 25.1 Building the bow

237-0600 or (310) 328-9560 to find a local dealer—mine sells the eyepiece for $189.

Those with more limited resources can use a less direct method to measure angular size. Center a telescopic sight on the coma and let the earth's rotation carry the comet across the field of view. Rotate the sight's crosshairs so that the comet drifts straight along the horizontal line, then count how many seconds it takes the coma to pass completely across the vertical crosshair. If you know the comet's declination (from position measurements with the star atlas), the width of the coma in minutes of arc is simply one quarter the cosine of the declination times the number of seconds. Repeat the measurement at least three times and average the results.

With a small telescope and a little practice, you can also estimate the comet's brightness, or visual magnitude, by comparing it with stars of known magnitude. Put the comet in sharp focus using a low-magnification eyepiece (no more than 2× magnification per centimeter of telescope aperture) and commit its image to memory. Point your telescope to a nearby star of known magnitude and defocus the image until the star appears the same size as did the comet. Then mentally compare the brightness of the defocused star and the comet. Find one star just slightly dimmer than the comet and another just slightly brighter—recalling that smaller magnitudes mean brighter stars—and you should be able to estimate where the comet's brightness falls in the interval between them. For more information about the magnitude scale, consult any basic astronomy text.

There are a few cautions to observe when estimating magnitudes. The atmosphere absorbs much more light when a star—or comet—is close to the horizon, so if the comet is at an elevation of less than 30 degrees, compare it only with stars that are at about the same elevation. Don't use red stars for comparison, because your eyes aren't very sensitive to red. If your catalogue lists a star as type K, M, R or N, or if the listing for V-B (visible-minus-blue) magnitude exceeds 1.0, find a bluer star. You will probably find it useful to practice this technique by estimating the brightness of stars of known magnitude. Experienced observers can achieve a precision of 0.1 or 0.2 magnitude.

To find out more about observing comets or to learn how to contribute your observations, contact the Harvard-Smithsonian Center for Astrophysics at icq@cfa.harvard.edu, or visit their World Wide Web site at *http://cfa-www.harvard.edu/cfa/ps/icq.html* or write to Daniel W. E. Green, Smithsonian Astrophysical Observatory, 60 Garden St., Cambridge, MA 02138. I gratefully acknowledge informative conversations with Dan Green. You can purchase the center's *Guide to Observing Comets*, the definitive resource on the subject, by sending $15, payable to International Comet Quarterly, to the same address. And do contribute your observations. Information is useless if it is not shared.

26 A PICTURE-PERFECT COMET

By Shawn Carlson, February 1997

I spent my formative years in what was then rural San Diego County. Our front porch opened onto an expanse of foothills so wild that we often startled coyotes away from our front yard when we retrieved the morning paper. The night sky was absolutely dazzling, and the millions of stars overhead enticed me out to enjoy them in secret.

Throughout the summer that separated second from third grade, my head never hit the pillow before I was scheming that night's escape. I had recently heard about Halley's comet—of its famous appearance of 1910 and of its imminent return. So, armed with a noisy party favor to scare off the coyotes, I posted scores of predawn hours atop a nearby hill, fully expecting to be the first person to see the comet on its return journey. By the time autumn's chill finally drove me off my hill, I felt utterly defeated. Only years later did I discover that I had misheard the year of Halley's return as 1968 instead of 1986.

In the decades since that disappointment, I've continued my love affair with the night sky, but I've never seen a cometary display to match the description of Halley's 1910 appearance. Halley's pathetic showing in 1986 left that stiff-lipped little boy within me still waiting for the heavens to redeem themselves.

Well, in 1997 they finally did. Comet Hale-Bopp (C/1995 O1) provided one of the most dazzling astronomical displays in modern times. The comet's nucleus is at least three times larger than Comet Halley's, and it ejected bright jets of dust for many months.

Even though Hale Bopp is long gone, it's important to remember that comets are frequent occurrences in the solar system. Several new comets are discovered every year. Bright comets, however, are rare.

This article explains how you can make a detailed photographic record of the next one. For low-tech ways of visually tracking a comet see Chapter 25, "Catch a Comet by Its Tail."

If you think that these days so many pictures are going to be taken of any comet that yours can't be scientifically useful, think again. Professional astronomers want your help. A comet's tail changes rapidly, and the professionals need an army of observers to track these changes. For example, every few days, when the magnetic field created by the solar wind changes direction, the gas tail can separate from the comet's head. The comet can repair these "disconnection events" in just 30 minutes; you could be the only person who records it.

With a good 35-millimeter camera (single-lens reflex), a few lens attachments and a little practice, you can take vivid comet portraits. Manual cameras are better for this work than the newer electronic automatic models. They perform better in cold weather, and you never risk a power failure during a long exposure. Try to find one that will let you lock up the internal mirror so that it won't jar the camera when you trip the shutter. If, while prowling around a used camera shop, you find a much coveted Olympus OM-1, Nikon F series or Pentax LX, buy it! Finally, because any vibration can spoil your image, you'll also need a cable release (available through camera supply companies for about $20) to activate the shutter without touching the camera.

When it comes to film, there is a trade-off between speed and grain size. Faster films require shorter exposures to catch the same detail, but they have a larger grain size and so give poorer resolution. As a general rule, you should use the slowest film your subject will allow, but nothing slower than ISO 400. Many astronomers prefer black-and-white film for its superior resolution. Kodak's T-Max 400 gives excellent results. If you decide to take color pictures, Fuji's Super G 800 and Kodak's Royal Gold 1000 get high marks for prints, and for slides Kodak's Ektachrome P1600 nudges out Fuji's Provia 1600 in side-by-side comparisons.

The correct exposure time depends on too many factors to guess accurately. The best advice is to shoot the comet at several exposures ranging from 10 seconds to 10 minutes. In general, a wide-angle lens requires shorter exposures than a telephoto lens of the same aperture, and the larger the aperture of the lens, the shorter the required exposure time. After reviewing the results of a roll or two, you should be able to narrow the range of exposure times.

The earth's rotation steadily shifts the sky, causing stars (and comets) to create curved streaks on any extended-exposure picture taken by a stationary camera. To compensate, you'll need to shift the camera with the sky throughout the exposure for any exposure longer than one minute. If you have a telescope with a sidereal drive, you can buy a piggyback mount to attack your camera to the telescope. (Lumicon in Livermore, Calif., sells one for $60; telephone: 510-447-9570.) Battery-operated drives for cameras are also available but very expensive. (Pocono Mountain Optics in Daleville, Pa., sells one for $295; telephone: 800-569-4323 or 717-842-1500.) Or you can build a hand guider for less than $20 [*see illustration on page 210*]. This design comes compliments of Dennis Mammana, resident astronomer of the Fleet Space Theater and Science Center in San Diego, and is the product of the collective cleverness of Dennis and other gifted astrophotographers. By manually twisting the adjusting screw one quarter turn every 15 seconds, you can eliminate star trails for up to 10 minutes on wide-field shots. Or use a high-torque DC motor powered by a car battery to turn the screw, then sit back and enjoy the comet.

Many astrophotographers take beautiful wide-angle portraits of comets that contrast the feathery tails against a stark landscape. Though often stunning, these images fail to capture many of the tail's most interesting features. Better science requires a closer view. The ideal field of view for recording tail features is about five degrees, which requires a lens with a 400-millimeter focal length. Because the tail may extend 20 degrees or more, you'll need to make a mosaic of images to catch the whole comet.

Remember, the longer the shutter is open, the more bad things can happen to your picture, so keep exposures as short as possible. This means you want a "fast"—that is, wide aperture—lens. The highest-quality 400-millimeter lenses can cost up to $6,000. Don't fret if you have to go to a shorter lens and a wider angle. A standard 135-millimeter telephoto is still scientifically useful. You will have to do some market research and balance your budget against the science you wish to accomplish. If possible, avoid mirror telephoto lenses and zoom lenses, whose optics are generally unsuitable for the sharp contrasts of astrophotography.

A few things to keep in mind: To help avoid darkroom catastrophes, begin each roll of film with a well-lit picture of something—anything. Without a reference image, darkroom technicians often can't identify the edges of a starry frame and sometimes miscut the film (to eliminate the risk entirely, many astrophotographers have their slides or negatives returned uncut). Also, carefully log the date, time and sky conditions for every observation you make. Most important, you must commit yourself to a regular observing schedule. A series of consistent observations made

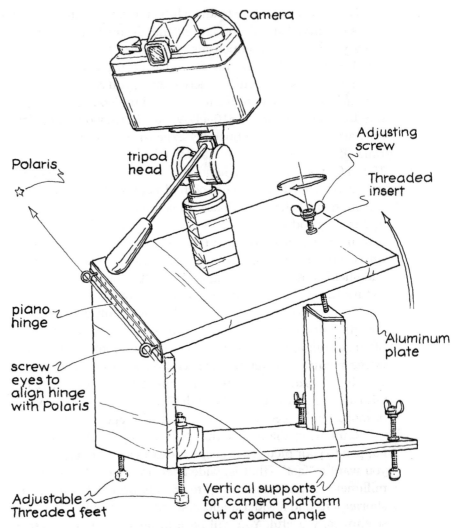

Camera

Adjusting
screw

Threaded
insert

Polaris

tripod
head

piano
hinge

screw
eyes to
align hinge
with Polaris

Aluminum
plate

Adjustable
Threaded feet

Vertical supports
for camera platform
cut at same angle

Figure 26.1   Hand guider for astrophotography

from the same site over many nights is much more valuable than a scattering of disjointed observations taken around the world. To do the job right, you've got to be out there on every clear night possible.

To find out more about observing comets or to learn how to contribute your observations, contact the Harvard-Smithsonian Center for Astrophysics via e-mail at icq@cfp.harvard.edu or visit their World Wide Web

Hand Guider for Astrophotography

The guider is essentially a hinged wedge that can hold a camera steady while rotating its field of view to keep up with the movement of the stars through the night sky. Its crucial components are a wooden wedge, cut so that its angle matches the latitude of your observing site; a piano hinge to align the camera platform with the wedge; and an adjusting screw that turns inside a threaded insert to move the platform. For the one-turn-per-minute rate given in the text, the adjusting screw must be one quarter inch in diameter, with a pitch of 20 threads per inch; its centerline must be located precisely 11¾ inches from the hinge line. The top of the block on which it bears should be cut to the same angle as the wedge and protected by a small aluminum plate so that the screw does not cut into the wood.

To use the guider, set it up on a smooth, level surface and turn it so that the line of the piano hinge points directly at the North Star. Without moving the platform, turn the camera so that its viewfinder captures your item of interest. Lock up the shutter and begin your exposure, turning the adjusting screw gently one quarter turn every 15 seconds.

site at *http://cfa-www.harvard.edu/cfa/ps/icq.html*. I gratefully acknowledge informative conversations with Daniel W. E. Green of the Smithsonian Astrophysical Observatory and Dennis Mammana.

Further Reading

Astrophotography II. Patrick Martinez. Willmann-Bell Publishers, 1987.

Guide to Observing Comets. Smithsonian Astrophysical Observatory. Available for $15 from *International Comet Quarterly*, 60 Garden St., Cambridge, MA 02138.

Astrophotography: An Introduction. H.J.P. Arnold. Sky Publishing, 1995.

27 A DEVICE TO SIMULATE PLANETARY ORBITS

Conducted by C. L. Stong, October 1958

For amateurs who prefer to study satellites indoors, David Berger, who lived in Croton-on-Hudson, NY, submitted this orbit simulator when he was 14.

"This simulator," he writes, "consists of a rubber sheet stretched over a circular hole in a piece of plywood, the rubber being depressed in the center by a short stick as shown in the accompanying illustration [*page 213*]. The depression forms a three-dimensional curve and represents the gravitational field in the vicinity of a body in space. Assuming that the edge of the depressed sheet is level, a freely rolling body such as a steel ball, when placed near the edge of the sheet, will roll toward the center of the sheet. In effect the ball is 'attracted' to the center by a 'force' which depends on the depth to which the center is depressed. If the freely rolling ball is brought near the edge and given a circular push, it will orbit around the depression as a 'planet.' The push simulates the kick given to a satellite by the final-stage rocket.

"The friction of rolling simulates the drag of the atmosphere on a satellite. The ball accordingly spirals closer to the center of the depression. As the average radius of the orbit decreases, the curve becomes tighter and the ball speeds up, just as a satellite does when it spirals closer to the earth to conserve its orbital angular momentum. One can also demonstrate the eccentric orbit of a comet and an orbit in which a body escapes.

"The effect of two bodies acting on a satellite in space can be simulated by inserting two sticks spaced a few inches apart under the supporting bridge. With this arrangement and variations of it a variety of orbits can be simulated: an earth-to-moon orbit in the form of a figure eight in which the satellite returns to earth, an S-shaped earth-to-moon orbit which terminates on the moon, and an earth-to-moon elliptical orbit which returns to earth."

Roger Hayward, who illustrates this department, admits to being something of a table-top-satellite enthusiast. "David Berger," he writes, "deserves a real pat on the back for calling the attention of high-school-physics teachers to this nice demonstration of celestial mechanics. In the interest of strict accuracy, however, the curve can be improved. The displacement of a point on a circular membrane of rubber which is loaded with a force at the center is equal to $(2F/T) \ln (a/r)$. F and T have to do with the forces applied and the character of the membrane, ln is the natural logarithm, a is the radius of the membrane and r is the radius of the point in question. This formula indicates that the force acting toward the center on a ball rolling on the curve would vary inversely with the radius. Such a force would give rise to elliptical orbits. But the 'attracting' body would be at the center of the ellipse instead of at one focal point, and thus would present a most uncelestial appearance.

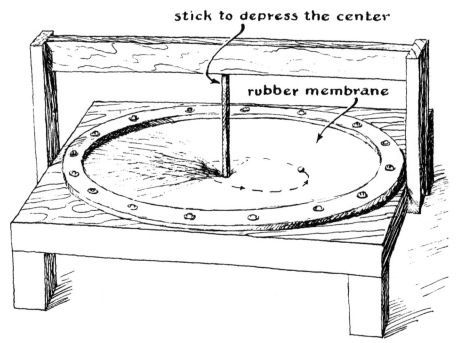

Figure 27.1 An amateur's device to simulate the orbits of artificial satellites

"The required surface has a profile of the form $x = 1/y$, the slope of which varies inversely as the square of the distance. The difference between this profile and that of the rubber membrane is shown in the accompanying illustration [*below*]. I turned such a surface in a disk of plaster with a lathe. The actual shape required is a petal curve which is removed from the $x = 1/y$ curve by a distance equal to the radius of the ball. Such a curve is difficult to compute but easy to make if the face of the cutter has the same radius as the ball. In making my surface I turned a series of rings in the plaster. The rings were spaced 1/10 inch apart. The compound rest of the lathe was used to set the depth of each cut, and the ridges between cuts were subsequently smoothed off by hand. Plaster cuts nicely but corrodes steel, so the lathe should be covered for protection.

"It is not always easy to prepare a plaster-cast free of bubbles. This difficulty can be overcome by proper mixing. Fill a container of adequate size with water and sprinkle the plaster into the center of the bowl by hand. This enables you to feel and remove any lumps. Plaster should be added

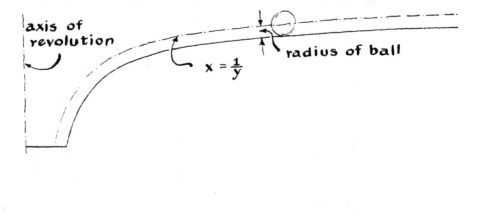

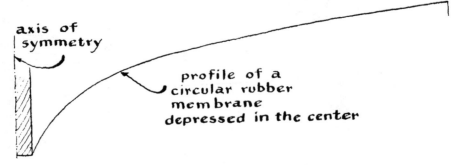

Figure 27.2 The profile of two curves used in orbit simulators

until the pile rises about an inch above the water in the center. The dry heap acts as an escape duct for trapped air. Allow about five minutes for the air to escape. Then the mixture may be stirred gently. Under no circumstances should the plaster be stirred before water has penetrated to the center of the pile, because the dry portion will be broken into small volumes from which air cannot escape.

"I cast the disk from which my surface was turned in a mold formed by wrapping a strip of gummed paper around a thick metal disk so that half the strip extended above the edge of the metal. The resulting cast was attached by sealing wax to a faceplate for machining. Never pat or trowel wet plaster or it will harden unevenly and increase the difficulty of accurate machining. Permit it to dry thoroughly. Wet plaster loads sandpaper quickly."

28 AN ASTROPHYSICAL LABORATORY IN YOUR BACKYARD

Conducted by C. L. Stong, September 1956

Strip the astronomical telescope of its clock drive, film magazine, spectrograph and related accessories and you put it in a class with a blind man's cane. Like the cane, it informs you that something is out in front. Shorn of appendages, the telescope tells you next to nothing about the size, temperature, density, composition or other physical facts of the bodies which populate space. Not more than 20 celestial objects, other than comets, appear through the eyepiece as interesting patterns of light and shade. Only one, the moon, displays any richness of surface detail. All other bodies look much as they do to the naked eye. There is a greater profusion of stars, but as a spectacle the night sky remains substantially unchanged.

That is why the experience of building a telescope leaves some amateurs with the feeling of having been cheated. A few turns at the eyepiece apparently exhaust the novelty of the show, and they turn to other avocations.

Other amateurs, like Walter J. Semerau of Kenmore, N.Y., are not so easily discouraged. They pursue their hobby until they arrive at the boundless realm of astrophysics. Here they may observe the explosion of a star, the slow rotation of a galaxy, the flaming prominences of the sun and many other events in the drama of the heavens.

Semerau invested more than 700 hours of labor in the construction of his first telescope. "I must confess," he writes, "that what I saw with it seemed poor compensation for the time and effort. That, however, overlooks other satisfactions: the solution of fascinating mechanical and opti-

216

cal problems. Considered in these terms, that first instrument was the buy of a lifetime."

Semerau soon decided, however, that he had to have a larger telescope equipped with devices to gather more information than his eye could detect. Accordingly he went to work on a 12½-inch Newtonian reflector, complete with film magazine and four-inch astrographic camera. Both were assembled on a heavy mounting with an electric drive, calibrated setting-circles and slow-motion adjustments. He could now not only probe more deeply into space but also do such things as determine the distance of a nearby star by measuring its change in position as the earth moves around the sun. To put it another way, he had made his "cane" longer and increased his control of it. When the sensitivity of modern photographic emulsions are taken into account, Semerau's new instruments were almost on a par with those in the world's best observatories at the beginning of the 20th century.

Since then, we have gained most of our knowledge of the physics of the universe. Most of this knowledge has come through the development of ingenious accessories for the telescope which sort out the complex waves radiated by celestial objects.

Semerau now decided that he had to tackle the construction of some of these accessories and to try his hand at the more sophisticated techniques of observing that went with them. He went to work on a monochromator, a device which artificially eclipses the sun and enables the observer to study the solar atmosphere. Semerau's description of the apparatus can be found in Chapter 13, "A Choronograph to View Solar Prominences."

Having built the monochromator, Semerau felt he was ready to attempt one of the most demanding jobs in optics: the making of a spectrograph. Directly or indirectly the spectrograph can function as a yardstick, speedometer, tachometer, balance, thermometer and chemical laboratory all in one. In addition, it enables the observer to study all kinds of magnetic and electrical effects.

In principle the instrument is relatively simple. Light falls on an optical element which separates its constituent wavelengths or colors in a fan-shaped array; the longest waves occupying one edge of the fan and the shortest the other. The element responsible for the separation may be either a prism or a diffraction grating: a surface ruled with many straight and evenly spaced lines. The spectrograph is improved by equipping it with a system of lenses (or a concave mirror) to concentrate the light, and with an aperture in the form of a thin slit. When the dispersed rays of white light are brought to focus on a screen, such as a piece of white cardboard, the slit appears as a series of multiple images so closely spaced that a continuous ribbon of color is formed which runs the gamut of the rainbow.

Each atom and molecule, when sufficiently energized, emits a series of light waves of characteristic length. These appear as bright lines in the spectrum and enable the investigator to identify the chemical elements of the incandescent source. Similarly, the atoms of a gas at lower temperature than the source absorb energy at these characteristic wavelengths from light transmitted through the gas. The absorption pattern appears as dark lines. As the temperature of the source increases, waves of shorter and shorter length join the emission, and the spectrum becomes more intense toward the blue end. Thus the spectral pattern can serve as an index of temperature.

The characteristic lines of a substance need not always appear at the same position in the spectrum. When a source of light is moving toward the observer, for example, its waves are shortened by the Doppler effect. In consequence the spectral lines of atoms moving toward the observer are shifted toward the blue end of the spectrum. The lines of atoms moving away are shifted toward the red. Velocity can thus be measured by observing the spectral shift.

When an atom is ionized, *i.e.*, electrically charged, it can be influenced by a magnetic field. Its spectral lines may then be split: the phenomenon known as the Zeeman effect. Intense electrical fields similarly leave their mark on the spectrum.

These and other variations in normal spectra provide the astrophysicist with clues to the nature of stars, nebulae, galaxies and the large-scale features of the universe. The amateur can hardly hope to compete with these observations, particularly those of faint objects. However, with well-built equipment one can come to grips with a rich variety of effects in the nearer and brighter ones.

"If you are willing to settle for the sun," writes Semerau, "you shuck off a lot of labor. A three-inch objective lens, or a mirror of similar size, will give you all the light you need. The rest is easy. Many amateurs have stayed away from spectroscopes because most conventional designs call for lathes and other facilities beyond reach of the basement workshop, and many are too heavy or unwieldy for backyard use.

"In 1952 I chanced on a design that seemed to fill the bill. My employer, the Linde Air Products Company, a division of the Union Carbide and Carbon Corporation, needed a special spectroscope for industrial research and could not find a commercial instrument that met their specifications. The Bausch & Lomb Optical Company finally located a design that looked promising. As things worked out, it was adopted and is now on the market. My instrument is a copy of that design.

"The concept was proposed by H. Ebert just before the turn of the century. The instrument is of the high dispersion, stigmatic type and employs a plane diffraction grating. As conceived by Ebert, the design was at least 50 years ahead of its time. In his day plane gratings were ruled on speculum metal, an alloy of 68 per cent copper and 32 per cent tin which is subject to tarnishing. This fact alone made the idea impractical. Ebert also specified a spherical mirror for collimating and imaging the light. Prior to 1900 mirrors were also made of speculum metal. It was possible but not practical to repolish the mirror but neither possible nor practical to refinish the finely ruled grating. Consequently a brilliant idea lay fallow, waiting for someone to develop a method of depositing a thin film of metal onto glass that would reflect light effectively and resist tarnishing. Then John Strong, director of the Laboratory of Astrophysics and Physical Meteorology at the Johns Hopkins University, perfected a method of depositing a thin film of aluminum on glass.

"The process opened the way for many new developments in the field of optics. One of these is the production of high-precision reflectance gratings ruled on aluminized glass. Prior to being coated the glass is ground and polished to a plane that does not depart from flatness by more than a 100,000th of an inch. The metallic film is then ruled with a series of straight, parallel saw-tooth grooves—as many as 30,000 per inch. The spacing between the rulings is uniform to within a few millionths of an inch; the angle of the saw-tooth walls, the so-called 'blaze angle,' is held similarly constant. The ruling operation is without question one of the most exacting mechanical processes known, and accounts for the high cost and limited production of gratings.

"In consequence few spectrographs were designed around gratings. Then in 1951, Bausch and Lomb introduced the 'certified precision grating.' These are casts taken from an original grating. It is misleading to describe them as replicas, because the term suggests the numerous unsatisfactory reproductions which have appeared in the past. The Bausch and Lomb casts perform astonishingly well at moderate temperatures and will not tarnish in a normal laboratory atmosphere. The grooves are as straight and evenly spaced as those of the original. The blaze angle can be readily controlled to concentrate the spectral energy into any desired region of the spectrum, making the gratings nearly as efficient for spectroscopic work as the glass prisms more commonly used in commercial instruments. Certified precision replicas sell at about a tenth the price of originals; they cost from $50 to $1,000, depending upon the size of the ruled area and the density of the rulings.

"The remaining parts of the Ebert spectrograph—mirror, cell, tube, slit and film holder—should cost no more than an eight-inch Newtonian reflector.

"There is nothing sacred about the design of the main tube and related mechanical parts. You can make the tube of plywood or go in for fancy aluminum castings, depending upon your pleasure and your fiscal policy. If the instrument is to be mounted alongside the telescope, however, weight becomes an important factor. The prime requirement is sufficient rigidity and strength to hold the optical elements in precise alignment. If the spectrograph is to be used for laboratory work such as the analysis of minerals, sheet steel may be used to good advantage. For astronomical work you are faced with the problem of balancing rigidity and lightness. Duralumin is a good compromise in many respects. Iron has long been a favored material for the structural parts of laboratory spectrographs because its coefficient of expansion closely approaches that of glass. When mirrors are made of Pyrex, an especially tough cast iron known as meehanite has been used to counteract the effects of temperature variation.

"The optical elements of my instrument are supported by a tube with a length of 45 inches and an inside diameter of 8¼ inches. The walls of the tube are a sixteenth of an inch thick. The eight-inch spherical mirror has a focal length of 45⅜ inches. The grating is two inches square; it is ruled with 15,000 lines per inch. The long face of the saw-tooth groove is slanted about 20 degrees to the plane of the grating. The width of each groove is 5,000 Angstrom units, or about 20 millionths of an inch. Such a grating will strongly reflect waves with a length of 10,000 A., which are in the infrared region. The grating is said to be 'blazed' for 10,000 A. A grating of this blazing will also reflect waves of 5,000 A., though less strongly. These waves give rise to 'second-order' spectra which lie in the center of the visible region: the green. In addition, some third-order spectra occur; their wavelength is about 3,300 A. Waves of this length lie in the ultraviolet region.

"The angle at which light is reflected from the grating depends upon the length of its waves. The long waves are bent more than the short ones; hence the long and short waves are dispersed. A grating blazed for 10,000 A. will disperse a 14.5-A. segment of the first-order spectrum over a millimeter. My instrument thus spreads a 2,200-A. band of the spectrum on a six-inch strip of film.

"The film holder of my spectroscope is designed for rolls of 35-millimeter film. Light is admitted to the holder through a rectangular port six inches long and four tenths of an inch wide. By moving the holder across the port, it is possible to make three narrow exposures on one strip. This is a convenience in arriving at the proper exposure. The exposure time is estimated on the basis of past experience for one portion of the film; the interval is then bracketed by doubling the exposure for the second portion and halving it for the third.

"The most difficult part of the spectrograph to make is the yoke which supports the grating. Much depends on how well this part functions. It must permit the grating to be rotated through 45 degrees to each side, and provide adjustments for aligning the grating with respect to the mirror. The ruled surface must be located precisely on the center line of the yoke axis, preferably with provision for tilting within the yoke so that the rulings can be made to parallel the axis. In my arrangement this adjustment is provided by two screws which act against opposing springs, as shown in the drawing on page 222. The pressure necessary to keep the grating in the parallel position is provided by four springs located behind it. Two leaf springs, one above the other, hold the grating in place. The assembly is supported by an end plate from which a shaft extends. The shaft turns in a pair of tapered roller-bearings which, together with their housing, were formerly part of an automobile water-pump. A flange at the outer end of the housing serves as the fixture for attaching the yoke assembly to the main tube. It is fastened in place by two sets of three screws each, the members of each set spaced over 120 degrees around the flange. One set passes through oversized holes in the flange and engages threads in the tube. These act as pull-downs. The other set engages threads in the flange and presses against the tube, providing push-up. Adjusting the two sets makes it possible to align the yoke axis with respect to the tube.

"The shaft of the yoke is driven by a single thread, 36-tooth worm gear that carries a dial graduated in one-degree steps. The worm engaging the gear also bears a dial, graduated in 100 parts, each representing a tenth of a degree. The arrangement is satisfactory for positioning spectra on the ground glass or film but is inadequate for determining wavelengths.

"All plane gratings should be illuminated with parallel rays. Hence the entrance slit and photographic plate must both lie in the focal plane of the mirror. Small departures from this ideal may be compensated by moving the mirror slightly up or down the tube.

"The spectral lines of the Ebert spectrograph are vertical only near the zero order and tilt increasingly as the grating is rotated to bring the higher orders under observation. The tilting may be compensated by rotating the entrance slit in the opposite direction while viewing the lines on a ground glass or through the eyepiece. The effect is aggravated in instruments of short focal length.

"The cell supporting the mirror, and its essential adjustments, are identical with those of conventional reflecting telescopes. If no cell is pro-

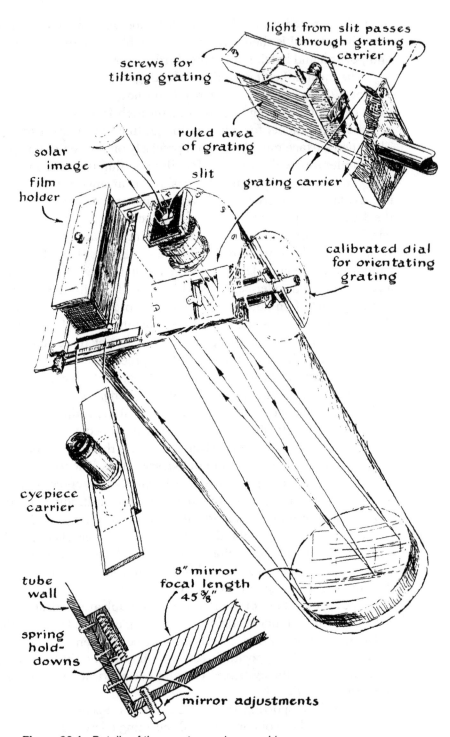

Figure 28.1 Details of the spectrograph assembly

vided and the adjustment screws bear directly on the mirror—which invites a chipped back—then no more than three screws, spaced 120 degrees apart, should be used. This is particularly important if the screws are opposed by compression springs; more than three will almost certainly result in a twisted mirror.

"The film magazine is equipped with a 48-pitch rack and pinion, purposely adjusted to a tight mesh so each tooth can be felt as it comes into engagement. It is this arrangement that makes it possible to move the film along the exposure port and make three exposures on each strip of film. Lateral spacing during the racking operation is determined by counting the meshes. Although the magazine accommodates standard casettes for 35-mm. film, it is not equipped with a device for counting exposures. I merely count the number of turns of the film spool and record them in a notebook.

"The back of the magazine is provided with a removable cover so that a ground glass may be inserted as desired. It also takes a 35-mm. camera, a convenience when interest is confined to a narrow region of the spectrum such as the H and K lines of calcium or the alpha line of hydrogen. The back may be changed over to an eyepiece fixture which may be slid along the full six inches of spectrum. This arrangement provides for a visual check prior to making an exposure; it is especially helpful to the beginner.

"Care must be taken in illuminating the slit. If the spectrograph has a focal ratio of $f/20$ (the focal length of mirror divided by the effective diameter of grating), the cone of incoming rays should also approximate $f/20$ and the axis of the cone should parallel the axis of the mirror. The slit acts much like the aperture of a pinhole camera. Consequently, if the rays of the illuminating cone converge at a greater angle than the focal ratio of the system, say $f/10$, they will fill an area in the plane of the grating considerably larger than the area of the rulings. Light thus scattered will result in fogged film and reduced contrast. Misalignment of the incoming rays will have the same effect, though perhaps it is less pronounced. Baffles or diaphragms spaced every three or four inches through the full length of the tube will greatly reduce the effects of stray light, such as that which enters the slit at a skew angle and bounces off the back of the grating onto the film. The diaphragms must be carefully designed, however, or they may vignette the film.

"The components are assembled as shown in the drawing on page 222. The initial adjustments and alignment of the optical elements can be made on a workbench. An electric arc using carbons enriched with iron, or a strong spark discharge between iron electrodes, makes a convenient source of light for testing. The emission spectra of iron have been determined with great precision, and the wavelengths of hundreds of lines extending far into the ultraviolet and infrared (from 294 to 26,000 A.) are tabulated in standard reference texts. Beginners may prefer a mercury arc or glow lamp because these sources demand less attention during operation and emit fewer spectral lines which are, in consequence, easier to identify. The tabulations, whether of iron or mercury, are useful for assessing the initial performance of the instrument and invaluable for calibrating comparison spectra during its subsequent use.

"Recently I have been concentrating on the spectroscopic study of sunspots. To make a spectrogram of a sunspot you align the telescope so that the image of the sun falls on the entrance slit. The objective lens of my telescope yields an image considerably larger than the slit. The image is maneuvered, by means of the telescope's slow-motion controls, until a selected sunspot is centered on the slit, a trick easily mastered with a little practice. The spectrum is then examined by means of either the eyepiece or the ground glass. The spot is seen as a narrow streak which extends from one end of the spectrum to the other. The adjustments, including the width of the entrance slit, are then touched up so the lines appear with maximum sharpness.

"Successive spectral orders are brought into view by rotating the grating through higher angles. Shifting the grating for the detection of a higher order is analogous to substituting eyepieces of higher power in a telescope. You get a bigger but proportionately fuzzier picture. The film magazine is substituted for the eyepiece and three exposures made in both the first and the second order. In many cases the range of intensity between the faintest and brightest lines exceeds the capacity of the film to register contrast. Three exposures, one estimated for the mid-range intensity and the other two timed respectively at half and twice this value, will usually span the full range.

"Gases in the vicinity of a sunspot often appear to be in a state of violent turbulence. At any instant some atoms are rushing toward the observer and others away. The spectral lines show proportionate displacement from their normal positions in the spectrum—the Doppler effect—and register as a bulge in the central part of the line occupied by the sunspot."

29 AN OCULAR SPECTROSCOPE

Conducted by Albert G. Ingalls, December 1952

As early as 1814 Joseph von Fraunhofer, the father of astrophysics, placed a prism before the 1.2-inch lens of a theodolite and mapped the dark lines of the solar spectrum he saw, designating them with the now familiar letters. These are the Fraunhofer lines that give the stars the separate individualities of different human faces—individualities that are but dimly realized by those who observe only with a telescope. Unlike the telescope, the spectroscope reaches into a star and takes a sample of it. Paul W. Merrill of the Mount Wilson and Palomar Observatories has said that studying a star by telescope is like "trying to guess the contents of a book from its size, weight and general appearance; while a spectroscopic observation is opening the book and reading it through line by line."

Today astrophysics, which deals with the physical and chemical characteristics of the stars, is the largest branch of astronomy; in fact, the astrophysical tail now wags the astronomical dog. Yet not one amateur astronomer in 100 uses even a simple spectroscope or seeks to become an amateur astrophysicist. True, much of astrophysics is abstruse, but not all of it. Getting started has been the chief obstacle.

A simple way to get a start in astrophysics is to build the little ocular spectroscope described by Roger Hayward's drawing on the next page. With it you can study the spectra of the brightest stars, including the sun, directly or as reflected by the moon. This spectroscope will show the more prominent lines of the solar spectrum when held in the hand and aimed at the sun. But when you insert it in the telescope in place of the eyepiece,

225

take care not to look through it directly at the sun, for that can make you blind. Without the telescope the spectroscope may also be used on light sources such as neon tubes, a salted gas flame or a welder's iron arc.

It is called an ocular spectroscope because its diameter is uniform with the standard telescope ocular, or eyepiece. It is kept with the set of oculars and adds variety to their use. Its multicolored diffraction-grating spectra will also serve to satisfy the astronomically unsophisticated visitors whom all telescope owners occasionally have to entertain and who, seeing only with their eyes and not with their understanding, fail to be impressed. The ocular spectroscope will make your Aunt Emma say "Ah!" even though she may never have heard of Kirchhoff's three laws of spectrum analysis.

The midget spectroscope was designed and built by Ernst Keil, an amateur astronomer and professional instrument maker at the California Institute of Technology in Pasadena, Calif. As an avocation he has from time to time designed and built little ocular spectroscopes, including one for James Fassero, the author of *Photographic Giants of Palomar*, who uses it in his lecture demonstrations with the 100-inch telescope at Mount Wilson.

The achromatic lens of about two inches focal length may be obtained from Edmund Industrial Optics for about $50 (*www.edmundscientific.com*), or a plano-convex lens may be substituted with little optical loss. The only working dimension is the 1¼-inch outside diameter, a carefully machined sliding fit for the telescope drawtube. The other dimensions are those you choose. There are no "blueprints." Transmission gratings are now photographically etched in ordinary film and sold in photographic slide holders. Edmund Industrial Optics (*www.edmundscientific.com*) sells such transmission gratings with 25,400 lines per inch in sets of two for under $10.

"The slit," Keil writes, "consists of two steel jaws made with care, their razor-sharp edges perfectly straight. The better the jaws, the sharper and more distinct will be

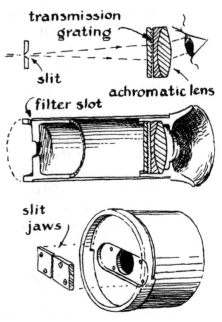

Figure 29.1 A simple ocular spectroscope

the spectrum lines. Their separation will depend upon the brightness of the star observed, but .01 inch should be suitable.

"The light from a star is gathered by your telescope and focused in the plane of the slit jaws. Entering the slit, it passes through the transmission grating, which disperses it into its colors, then through a lens that collimates the light (making it parallel) and magnifies the spectrum. In this spectroscope the grating is put behind the collimator, instead of in front of it, to protect the grating. Actual trial will prove that in this simple spectroscope it makes no difference on which side of the grating the collimator is placed, for the spectroscope is not intended for serious scientific research but only for demonstrating the elementary principles of spectroscopy.

"To put the instrument in operation, first rotate the grating-lens unit, which must have a sliding fit inside the outer tube, until the grating lines are parallel with the slit. Then slide it in or out until the slit is in sharp focus. Insert it in the telescope and move it in or out until brilliant spectra appear.

"The slot on the front of the spectroscope is at right angles to the slit and of such a depth that a filter placed in it will cover one half of the slit. Two spectra, one above the other, will then be seen simultaneously—one the original, the other an absorption spectrum. Gelatin filters may be had from the Eastman Kodak Company (*www.kodak.com*) or you can use red or blue cellophane, obtainable at photography stores."

30 MONITORING VARIABLE STARS

Conducted by Albert G. Ingalls, February 1954

The next time you visit the station of an advanced radio ham take a look around the basement and back yard. The chances are better than even that you will find a well-made telescope. Many hams have two avocations, astronomy and electronics.

Apollo Taleporos, an avid ham operator, explains: "Astronomy has borrowed so many of radio's techniques, and the aims of electronics have become so closely identified with those of astrophysical research, that you have to stop and think today before you can decide in which field you are working. I enjoy physics, particularly the electromagnetic part of it. So I do a lot of playing around in both electronics and optics. The only thing that keeps me out of the ultra-ultrashort waves is the cost of cyclotrons and sounding balloons."

Until 1946, according to Taleporos, most radio hams stuck closely to the communication side of radio. Back in 1912, when a government official was asked what part of the radio spectrum he would assign to the amateurs, he replied: "We'll stick 'em on 200 meters and below; they'll never get out of their back yards with that." As history records, in little more than two decades the hams made a back yard of the whole world, one now largely occupied by the multi-billion-dollar communications industry. But by the end of World War II amateur radio as such held few remaining challenges and the hams were ripe for something new.

It came suddenly, not with a bang but a squeal. Shortly after the Federal Communications Commission lifted the wartime ban on ham activity, some of the amateurs began to hear eerie squeals on the 14-megacycle band. The squeals were heard only at night, after the band had closed down for long-distance communication. An amateur would be idly scanning a portion of his dial between two active local stations in the hope of making one last distance contact before turning in for the night when suddenly a strange "woweeee" would be heard. It was unlike anything heard on the longer waves. Some evenings it would come a dozen or more times at the same point on the dial. On others it would not be heard at all.

Word passed around quickly and before long almost everyone with a supersensitive receiver equipped for tuning in continuous wave telegraph signals was listening to the mysterious wails. Then someone discovered that they came from meteors.

Meteors leave in their wake ionization trails—mile-thick cylinders of charged gas which reflect 14-megacycle waves efficiently. During the brief interval that a meteor's trail persists, a favorably located receiver can pick up the carrier wave of a transmitter which would otherwise be beyond range, for at night the ionosphere loses some of its charge and becomes less efficient in reflecting signals. Thus in effect the hams were hearing shooting stars. The discovery opened the era of radio astronomy for them.

The flood of surplus optical and electronic gear that was put on the market after the war was a big help. Much of it was sold "as is" in the form of complete units, and amateurs often had to buy a large piece of apparatus in order to get the particular meter, lens or gear train they wanted. So the whole thing was carted home and the useless parts stored away against the day when they might come in handy. Some of this "junk" consisted of parabolic antennas, photomultipliers, recording meters, prisms, filters and other odds and ends which were destined—with the help of sweat and imagination—to become priceless possessions.

Today many hams are seen more often than they are heard. Amateurs are busy tuning in celestial objects on radio telescopes, bouncing waves off the moon, tracking meteors and recording their paths automatically, detecting and measuring disruptions on the sun's surface, plotting the orbits of eclipsing binaries and measuring the apparent magnitudes of variable stars with a speed and precision which would have startled professionals in the days when Harlow Shapley suggested that the cepheid variable might make a good yardstick.

Just as meteor observing attracted the radio hams to astronomy, so the photoelectric cell made electronic addicts of many optical men. Since 1911 members of the American Association of Variable Star Observers have been looking at these stars through their small telescopes. They have made more than three and a half million observations of some 600 variable stars. But the job takes good eyesight. Now electronics can do it much better. With the variety of sensing tubes now on hand, electronic gadgets can see farther, faster and more sharply and keep at it longer than the human eye. Moreover, electronic eyes can "see" in invisible parts of the electromagnetic spectrum and can tolerate blinding intensities.

One of the first amateur astronomers to go all out for electronics was John Ruiz of Dannemora, N.Y. After several years of mixing vacuum tubes and lenses, he said, with only slight exaggeration: "Whatever you can do mechanically you can do better and cheaper electronically." Whereas in mechanical amplification accuracy to one part in 100,000 is considered very good, in electronic amplification accuracy to one part in 100 million is routine. It takes high craftsmanship to make a telescope objective that yields a magnification of 300 diameters with good resolution, but a beginner can build on his first try an electronic gadget capable of amplifying a million million times with equally good "resolution."

Electronics has also taken a load off the amateur astronomer's pocketbook. For example, an amateur can inexpensively—for under $100—pick up from the National Bureau of Standards' Station WWV time signals accurate to two parts in 100 million. How much would a pendulum clock of comparable accuracy cost, assuming, of course, that the market afforded one? The effect of the merger between astronomy and electronics has been to bring the amateur of modest means to a more nearly equal footing with his professional colleague in equipment.

A photoelectric photometer which Ruiz built for taking the guesswork out of variable star observing illustrates how simply optics and electronics can be combined to make an instrument of extraordinary power. As Ruiz says, you can hitch a photometer of your own make to a star.

"Of course, the star that you select should be one showing some action, *i.e.*, a variable star. Some variable stars go through a complete cycle in the course of a few hours (*e.g.*, the Beta Canis Majoris type), and you can make a full set of measurements for a complete 'light curve' in one night, weather permitting. Other stars of the eclipsing type have periods of a few days or longer. These are more difficult to follow and may require many nights of observing for a complete light curve. In the northeastern U.S. a series of observations on a single star may span two years or more, because nights ideal for photometry are rare.

"As an example of what an amateur can do with relatively modest equipment consider the light curve of No. 12 Lacertae, a short-period star of the Beta Canis Majoris type [*see upper illustration on page 231*]. This curve was taken in the course of one night. Ten curves like it were taken in the summer and fall of 1951 and they proved of value to the Dutch astronomer C. de Jager in making a new determination of the 'beat period' of the light variation. This star exhibits fluctuations both in amplitude and period, or, as radio hams would say, it exhibits both AM and FM.

"Another type of variable is illustrated by the complete light curve of Mu Herculis, an eclipsing variable with a very short period—two days and one hour. The curves were taken in two colors, blue and yellow, corresponding roughly to the respective photographic and photovisual magnitudes. These curves promise to be of value in a more accurate determination of the orbital elements of this system.

"As some astronomers have pointed out, visual observation of variable stars is now outmoded, at least for those of short periods and narrow range. The photoelectric method is at least 10 times as accurate and far more objective, particularly in ascertaining the brightness of stars of different color or spectral class.

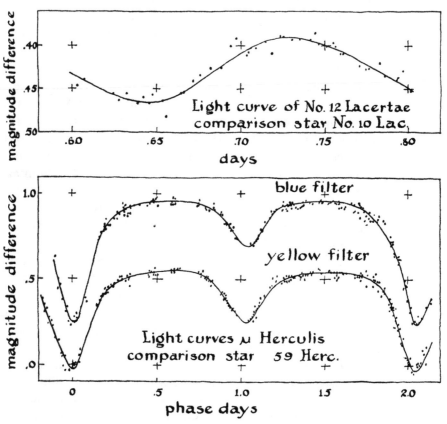

Figure 30.1 The light curves of two variable stars, as plotted by electronic photometer

"To make graphs like these, you must have, first of all, a telescope solidly mounted in the manner advocated by Russell Porter. My setup is shown in Roger Hayward's illustration [*see next page*]. The telescope should be provided with an accurate clock drive and slow motions in both declination and right ascension. A reflecting telescope is preferred to a refractor, because violet and near ultraviolet light, to which the cesium-antimony photocell is most sensitive, is reflected by the aluminum surface of a mirror more effectively than it is transmitted by a refractor. Incidentally, the prism of a Newtonian should be replaced by an aluminized flat for the same reason. For the conventional eyepiece and eye we substitute a photometer head, or light box, which consists mainly of a holder for filters and the photomultiplier tube.

"In principle the stellar photometer is little more than a glorified exposure meter of the type used in making photographs. Like *omnia Gallia,* it consists of three parts: the photometer head, which corresponds to the light cell of the exposure meter, a direct current amplifier to build up the faint energy received from the stars, and an indicating device, usually a milliameter, which is read by eye, if possible by the eye of an assistant.

"The construction of the photometer head calls for no special tools and is easy if the amateur can lay hands on a small junk camera. Mine was made according to suggestions from William A. Baum of the Mount Wilson and Palomar Observatories. It is built around a camera shutter provided with an iris diaphragm [*see inset in illustration on page 233*]. Note the use of the 'field lens,' which serves the double purpose of forming an enlarged image of the star (centered on the iris diaphragm) and of projecting on the photocathode an image of the fully illuminated mirror rather than a pinpoint image of the star under observation. This compensates for the granular structure of the photocathode, which is not equally sensitive at all points. You may compute the focal length of this field lens to give an image of your object glass or mirror about five millimeters in diameter, using nothing more than the elementary optics you learned in high school. Note that the gate holding the multiplier can be swung to one side when centering your star on the diaphragm. It is provided with an automatic shutter to cut off all light to the tube when this is done, for if strong light strikes the energized photocathode, it may become temporarily unstable or even permanently damaged.

"My photometer uses a nine-stage 1P21 photomultiplier which works on the secondary emission principle. Each stage amplifies the signal by a factor of four or five, so that the over-all gain in signal strength is enormous.

"The energy we receive from a faint star is so small that even after immense amplification by the dynode stages of the photomultiplier tube it

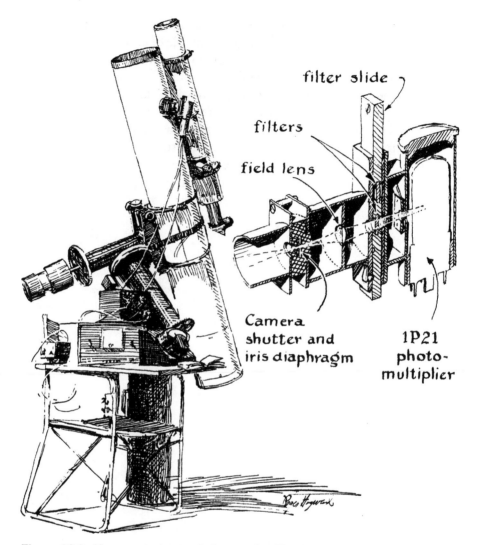

filter slide

filters

field lens

Camera
shutter and
iris diaphragm

1P21
photo-
multiplier

Figure 30.2 How an electronic photometer is affixed to a telescope, and a detail of
the photometer

may amount to no more than 10^{-9} ampere (nanoamp). Minute currents of
this order of magnitude can be measured by sensitive electrometers.

"With these separate units completed, all that remains is to hook them
together, substitute the photometer head for the eyepiece, select a star and
have fun. If you have both a persuasive personality and a kindly disposed
partner, one of you can take readings while the other guides the telescope
with the auxiliary telescope. As is evident, stellar photometry is an ideal

project for two amateurs, one a gadgeteer to build the apparatus and the other a researcher who enjoys the role of chief observer.

"To meet the present high standards of professional observation, an amateur must apply certain corrections which were of no consequence in the era of visual work. These have to do with atmospheric extinction, reduction of time to the sun and non-linearity of the meter and amplifier. But a serious worker will find many a helping hand in amateur and professional circles. The dividing line between a professional and an amateur is very dim indeed, for only an 'amateur' would ever become a professional. James Stokley of the General Electric Company once said that the difference between an amateur astronomer and a professional is that the amateur is sorry when it is cloudy."

GLOSSARY

Absolute magnitude: A measure of the inherent brightness of a celestial object. This scale is defined as the apparent magnitude a star would have if it were seen from a standard distance of 32.6 light-years (10 parsecs).

Achromatic lens: Achromatic lenses are specially fabricated to reduce the amount of chromatic distortion they generate. They generally consist of two or more optical elements. A doublet lens, having two elements, can focus two different wavelengths to the same location. A triplet lens can do the same for three wavelengths.

Alloy: An alloy is a mixture of different metals or a metal and something else.

Altazimuth: A mount in which the telescope swings in azimuth about the vertical direction and in altitude about the horizontal direction.

Aluminized mirror: A glass mirror that has been coated with a thin layer of reflective aluminum.

Ampere: An ampere (abbr. "amp") standard unit by which one measures electrical current. One ampere is defined to be the current that flows when one coulomb of charge moves past a given point in one second.

Angstrom: An angstrom is a unit of distance that is convenient for measuring atoms, atomic structures and the wavelengths of visible light. One angstrom equals 10^{-10} meters. An atom is about one angstrom in diameter. The wavelength of visible light ranges between about 4000 and 7000 angstroms.

Apparent magnitude: A measure of the brightness of a celestial object as seen from the earth. It is a logarithmic scale—the lower the number, the brighter the object. For example, if one object is ten times as bright as another, it will be 2.5 units less on the magnitude scale. Negative numbers indicate extreme brightness. The full moon has an apparent magnitude of −12.6; the sun's is −26.8. We can see objects up to 6th magnitude without a telescope.

Arc-minute: An arc minute is an angular measure of an arc, equal to one-sixtieth of a degree in a circle that has been subdivided into 360 degrees, or to 60 seconds of an arc.

Arc-second: An arc second is an angular measure of an arc, equal to one-sixtieth (1/60) of an arc minute.

Astigmatism: One kind of distortion that can be built into an image. Astigmatism is created by an optical element (a lens or a mirror) with a distortion of its surface that is not symmetric about the optical axis.

Atmospheric extinction: The property of the atmosphere by which it absorbs (extincts) light. This is critical for determining the magnitude of a given star. The amount of extinction depends on the height of the star in the sky, as well as the amount of haze and high altitude moisture, and a host of other factors.

Atmospheric dispersion: The property of the atmosphere by which it disperses (scatters) light.

Atmospheric refraction: The property of the atmosphere by which it refracts (bends) light. This can be quite significant when tracking stars at low altitudes, where the thickness of the atmosphere is greatest.

Azimuth: The horizontal angular distance from a reference direction. Usually the northern point of the horizon to the point where a vertical circle through a celestial body intersects the horizon, usually measured clockwise.

Barlow lens: An arrangement of lenses that artificially lengthens the focal length of the primary lens or mirror; by doing so the barlow lens increases the magnifying power of the instrument.

Capacitance: A measure of an object's ability to hold an electric charge. It is large when it takes very little energy to deposit a large charge on the object. An object's capacitance is defined to be the deposited charge divided by the voltage that appears when that charge is applied.

Capacitor: An electric circuit element used to store charge, usually consisting of two metallic plates separated and insulated from each other by a dielectric.

Carborundum: A trademark used for an abrasive of silicon carbide crystals. It's often used as a course grinding agent in shaping glass mirrors.

Cassegrain telescope: A general classification for reflecting telescopes in which the light reflects from the primary mirror to a secondary mirror, then back through a central opening in the primary mirror. The image is brought into focus a short distance behind it by an eyepiece.

Cassini division: The main dark gap between the largest rings of Saturn; discovered by Gian Domenico Cassini in 1675.

Catadioptric telescope: A wide-field telescope that uses a meniscus lens in front of a spherical primary mirror. The lens adds a negative spherical aberration that cancels the spherical aberration of the mirror.

CCD (charged coupled device) camera: Light can induce an electric charge on a piece of prepared silicon wherever it happens to strike. A CCD camera uses this effect to reconstruct an image; highly charged sections of the silicon wafer represent bright parts of the scene, areas of low charge represent dark regions. These images are convenient for astronomy, in part because they can be read out into a computer and processed electronically.

Celestial equator: The projection of the earth's equator onto the celestial sphere.

Chromosphere: The thin layer of the solar atmosphere that lies above the photosphere and below the corona. Most of the light that is produced in the chromosphere is emitted by hydrogen through the Hydrogen-alpha transition.

Coaxial cable: A type of signal cable that consists of a center wire surrounded by insulation and a grounded shield of braided wire. The shield minimizes electrical and radio frequency interference on a signal carried on the inner wire.

Coma: The luminescent cloud containing the nucleus and the gases making up the major portion of the head of a comet. Also, a type of distortion that can occur in optical systems that can degrade the quality of the image.

Comparison star: A star used as a standard on the magnitude scale.

Corona: The outermost and hottest region of the sun's atmosphere. The corona extends for millions of kilometers beyond the apparent surface of the sun and its temperature is actually much hotter than the sun's surface. In some places the temperature reaches millions of degrees C. The corona is visible to the naked eye, but only during a total solar eclipse.

Coronagraph: A device used to observe the sun's corona. It contains an occulting disk that is used to artificially obstruct the sun.

Correcting plate: A thin lens used to correct incoming light rays in special forms of reflecting telescopes. The lens is generally irregular in shape and is designed to reduce the effects of imperfections in the rest of the instrument's optics.

Chromatic aberration: A distortion of a multi-color image caused by the fact that rays of different colors (wavelengths) experience a different amount of refraction when they pass through a lens. As a result, rays of different wavelengths which pass through a lens are invariably focused to different points along the optical axis. This effect blurs the image.

Special lenses, called achromats, are specially designed to reduce this effect.

Constant current transformer: A device that converts one constant current into a constant current with a different value.

Cross-hairs: The cross pattern that one sees when looking through an eyepiece, used in order to center an object. The best material to make cross-hairs out of these days is carbon fiber which you can purchase at most hobby stores.

Crown glass: A soda-lime optical glass that is exceptionally hard and clear, with low refraction and low dispersion.

Dall-Kirkham telescope: A reflecting telescope in which the primary mirror has a spherical curvature and the secondary mirror has an elliptical curvature.

Decibel: A unit used to express the relative difference in power or intensity; equal to ten times the base-ten logarithm of the ratio of the two levels. When used to measure relative voltage, rather than energy, the equation is twenty times the base-ten logarithm of the ratio of the two voltages.

Declination: Declination is one of the coordinates that locates an object on the celestial sphere. It measures the angular distance of a celestial object north or south of the Celestial Equator much like latitude does on the surface of the earth. In astronomy, declinations in the Northern Celestial Hemisphere are positive. Southern hemisphere declinations are negative.

Diagonal mirror: The secondary mirror in a Newtonian telescope.

Dielectric: Any electrical insulator.

Diffracting prism: A prism used to separate light waves by wavelength.

Diffraction: Diffraction describes the change in the directions and intensities of a group of waves after passing by an obstacle or through an aperture.

Diffraction grating: A diffraction grating is a surface (usually made of glass or metal) with a large number of very fine parallel grooves or slits cut in the surface. It's used to produce optical spectra by diffraction of reflected or transmitted light.

Diode: An electronic device that allows current flow almost exclusively in one direction.

Dipole: An antenna, usually fed from the center, consisting of two equal rods extending outward in a straight line. Surging a sinusoidal current into the antenna causes a radiation pattern known as "dipole radiation."

Dobsonian mount: A very stable alt-azimuth mount with huge bearings.

The Dobsonian mount is very easy to build and so easy to manipulate by hand that it is a favorite of amateur astronomers.

Doublet lens: A lens made up of two single lenses. Often, it refers to two plano-convex lenses whose focal lengths are in the ratio of three to one, placed with their plane sides toward the object and the lens of shortest focal length next to the object.

Dry ice: Frozen carbon dioxide, temperature of −78 C. This material is widely available at welder's supply stores. Dry ice is often used to chill equipment when room temperature is high enough to degrade said equipment's performance.

Eclipse: The partial or complete obscuring of one celestial body by another relative to an observer.

Eclipse, Lunar: Obstruction of the view of the moon as seen by an observer on earth when the earth passes between the moon and the sun.

Eclipse, Solar: Obstruction of the view of the sun as seen by an observer on earth when the moon passes between the earth and the sun.

Eclipse, Star (more commonly called an **Occultation**): Obstruction of the view of a star by the moon or a planet.

Eclipsing variable: A variable star that passes behind a companion star and so has its signal obscured from our view.

Elasticity: A measure of the amount by which a material can be deformed and then return to its original shape.

Ellipse: All the points for which the sum of the distances to two fixed points is equal. One of the class of curves called a conic section which can be generated as the intersection between a cone and a plane. An ellipse has eccentricity less than one.

Elliptical mirror: A mirror with an elliptical curvature.

Equatorial mount: Telescope mount with one axis parallel to the rotational axis of the earth (Polaris in the Northern hemisphere, sigma Octantis in the Southern hemisphere) allowing one to follow an object by just counteracting the rotation of the earth.

Erecting telescope: A telescope that creates an erect (upright or true) image. Telescopes normally produce an inverted image. Erecting telescopes generally have an addition optical element to flip the image over and thus make it erect.

Erfle eyepiece: An eyepiece that consists of five or six separate elements with the lens closest to and farthest from the eye being achromatic doublets, with either a single or a doublet lens in between. These offer a wide field of view, extremely low distortion and outstanding color correction. They are also amongst the most expensive eyepieces you can buy.

F-ratio or **focal ratio** or **f/#:** The f- or focal ratio is the ratio of the focal length to the diameter of a telescope mirror. For a telescope, the focal ratio determines the magnification and thus the field of view.

f-stop: Photographers define the ratio of the focal length to diameter of their optical systems as the f-stop. A photographer can change the f-stop by adjusting the width of the iris covering the camera lens. This adjusts the amount of light that enters the camera without affecting the magnification. Controlling the f-stop allows a photographer to get a perfect exposure for a given set of conditions, like brightness of the scene to be photographed, focal length of the camera, speed of film and shutter speed.

Field of view or **image field:** This is the solid angle in which you can see things through an optical instrument (telescope, binocular, photographic camera, etc.).

Filter: A device that reduces the intensity of the light that shines through it.

Finder or **finderscope:** A small auxiliary telescope that mounts on your main telescope. It is used to help find objects to observe through the main telescope.

First surface mirror: A mirror with the reflective coating on the front surface so that light reflects from the surface without passing through the glass.

Flint glass: A very hard, fine-grained quartz that sparks when struck with steel.

Focal length: The distance from the surface of a lens or mirror to its focal point.

Focal plane: The plane onto which an image is focused by some optical system of lenses and/or mirrors.

Focus or **focal point:** Consider a group of parallel light rays that enter an instrument. If the rays pass through a converging lens the focus is the point at which those rays converge. If they pass through a diverging lens, the focus is the point from which they appear to diverge.

Foucault mirror test: A test of optical surfaces, developed by Leon Foucault, a French scientist. The test uses reflection and geometrical optical principles to amplify shadows of defects on the mirror so that they are easily visible.

Gadgeteer: A person who enjoys building scientific instruments.

Gas-discharge tube: A glass or quartz tube in which a rarefied gas is subjected to a large voltage between two metal electrodes. The gas rapidly ionizes when the voltage exceeds a critical value. This causes the gas to become conducting and a current passes between the electrodes through a spark inside the tube.

Gaussian: The classic "bell-shaped" curve that appears frequently in statistics.

Guide star: A bright star in a star field; one which a telescope can fix to hold the field stationary for the observer.

Hastings triplet: A particular eyepiece configuration that converts an inverted image into an erect image. It is sometimes used in terrestrial telescopes.

Hertz: A unit of frequency equal to one cycle per second.

Horsepower: A unit of power equal to 745.7 watts or 33,000 foot-pounds per minute.

Huygenian eyepiece: A relatively inexpensive, low-quality eyepiece composed of two plano-convex lenses with both plano surfaces facing the eye. These eyepieces provide lateral color correction and diminished coma with a flat field. Not often used with a reticule because the image plane is between the two lenses.

Hydrogen-alpha transition: The transition of an electron from the first excited state of hydrogen to the ground state. When this takes place the electron's excess orbital energy is shed by the emission of a photon with a wavelength of 656.4 nanometers.

Hyperbola: The group of points for which the difference of the distances from two given points is a constant.

Image-field: The visual field seen through a telescope.

Impedance, electrical: Impedance is a measure of the total opposition to current flow in an alternating current circuit, and is made up of two components: the ohmic resistance which does not depend on the signal's frequency, and the reactance which does. Reactance arises from a circuit's capacitance and inductance, because capacitors and inductors resist changes in their state. The greater the rate of charge (that is, the larger the signal's frequency), the greater is that resistance. Impedance can be represented in complex notation as $Z = R + iX$, where Z is the impedance, R is the ohmic resistance and X is the reactance.

Inductance: A changing magnetic field always generates a voltage. If a conductor is present, that voltage will drive a current which produces its own magnetic field that interacts with the changing field in such a manner as to always oppose the change. Inductance is defined as the negative of the induced voltage divided by the time rate of change of the current that produced the voltage. Inductance is a kind of inertia. It measures the ability of the circuit element to oppose a changing current.

Infrared light: The range of invisible light with wavelengths from about 750 nanometers to 1 millimeter.

Ion: A molecule with an excess charge. An ion may be created by removing electrons from or adding electrons to a neutral atom.

Ionize: The process of creating an ion from a neutral atom. Usually, this refers to removing an electron by adding energy to the atom's electron cloud.

Joule: A unit of energy equal to the work done when a force of 1 newton acts through a distance of 1 meter.

Kellner eyepiece: This eyepiece consists of an achromatic doublet near the eye and a plano-convex lens some distance behind with the flat side facing away from the observer's eye. The design is similar to the Ramsden, but has superior color correction due to the achromatic sub-unit.

Lap: see **Pitch lap.**

Laser: A device that produces a narrow beam of coherent light of a single wavelength by stimulating the emission of photons from atoms, molecules or ions.

Light curve: The curve generated by monitoring an object's brightness over time.

Light-tight: A term denoting a material or piece of equipment that is impenetrable by light.

Light-year: A unit of astronomical distance. It is the distance light travels in one year in a vacuum: about 5.87 trillion miles, or 9.43 trillion kilometers.

Liquid nitrogen: A coolant with a temperature of 77 K (equivalent to -321 F or -196 C) created by chilling nitrogen gas until it liquefies.

Magnification: Ratio of the angular size of an object to the angular size of the image of the object, created by a telescope (if the object is large but far away) or microscope (if the object is small and close).

Meniscus: The curved upper surface of a liquid in a container. The meniscus is concave if the liquid adheres to the container walls and convex if it does not.

Meniscus lens: A thin and highly curved lens.

Mirror power: See **focal ratio.**

Mount: A structure that supports any apparatus.

Negative lens: A simple concave lens that causes rays of light to diverge away from the optical axis.

Neutral density filter: A filter that reduces the intensity of light equally for all wavelengths.

Newton: The metric unit of force. One newton is the amount of force required to accelerate a mass of one kilogram at a rate of one meter per second per second.

Newtonian telescope: A reflecting telescope in which light reflects from the primary mirror to a diagonal mirror out to an eyepiece at the top of the telescope tube.

Non-linear: Any data that cannot be fit to a straight line.

Objective lens: Lens (or mirror) that receives the light from the object at which the telescope is pointed.

Occultation: Occultation occurs when one object becomes invisible because it passes behind another. It usually refers to a moon, a planet, or an asteroid. Heavenly bodies move in highly predictable cycles. Occulations often allow for precise calibration of their motions.

Optical flat mirror: A mirror that is flat across its surface to within one wavelength of visible light.

Oscillator: Anything with a physical state that repeats in a precise pattern. In electronics, an oscillator produces a repeating voltage signal that is usually a series square-wave pulse, but that may be sinusoidal.

Parabola: A plane curve formed by the group of points equidistant from a fixed line and a fixed point not on the line.

Parabolic antenna: A parabolic antenna is one in which a parabolic-shaped reflector is used to reflect radio waves to a receiver located at the focus. Parabolic antennas concentrate all radio waves at the same point, regardless of wavelength.

Parabolic mirror: A mirror with a parabolic curvature. For a parabolic mirror, all rays that come in parallel to the optical axis are reflected to the same point, called the focus.

Parametric amplifier: A device that amplifies by varying a parameter of a tuned circuit. An example would be varying the length of a pendulum at twice its natural frequency to amplify the motion. In electronics, one can vary the capacitance of a tuned circuit with a voltage variable capacitor by driving it with a "pump" signal. These devices are used for low-noise amplification.

Parsec: A unit of astronomical distance. It is the distance at which 1 AU would subtend an angle of 1 arc second. One parsec is equivalent to 3.26 light-years, or 19.2 trillion miles, or 30.7 trillion kilometers.

Pen recorder: An old-fashioned recording instrument in which the rise and fall of a signal is tracked by a pen that moves across a long strip of paper that is moved past the pen at a constant rate.

Persistence of vision: The effect by which an image can briefly be seen after light stops stimulating the retina.

Phase: The phase refers to a particular place in a wave's cycle.

Photocathode: A thin semiconductor coating of material that ejects electrons when struck by photons in or near the visible range.

Photocell: An electronic device that produces a current that varies in proportion to incident radiation.

Photoelectric measurement: A measurement made by using photons to trigger observable changes in an electronic system.

Photometer: An instrument for measuring light properties, especially luminous intensity or flux.

Photomultiplier tube: A device that is so sensitive to light that it can detect individual photons. Photomultiplier tubes can also resolve the time at which a photon arrives, typically to less than a few nanoseconds.

Photosphere: The photosphere is the innermost part of the solar atmosphere and defines the visible boundary of the sun.

Pitch lap: A pitch that acts as a cushion and is used for polishing lenses and mirrors.

Plasma: Gas that has been heated to the point at which random collisions between the gas molecules have enough energy to knock off some outer electrons so that the gas contains a stable concentration of ions.

Plasticity: The property by which a material may be deformed without suffering from a structural failure.

Plössl eyepiece: An eyepiece composed of two achromats with the crown glass sides facing each other. It provides extremely sharp color corrected images and a wide field of view.

Pocket telescope: A telescope that is designed to fit in the pocket. These telescopes generally have small focal ratios and large field of views.

Polar axis: The fixed reference axis from which the polar angle is measured in a polar coordinate system. For the earth, this refers to the axis of the earth's rotation.

Positive lens: A simple lens that causes light rays from a subject to converge to a point.

Potentiometer: A three-terminal resistor with an adjustable center connection that allows one to create a variable amount of resistance inside a circuit.

Primary mirror: Largest mirror, used as a converging lens in a reflecting telescope.

Prism: A transparent body usually with triangular ends, often used for separating white light passing through it into its spectrum. It is used for reflecting beams of light and rotating images.

Prominence: A filament that appears on the limb of the sun that extends above the chromosphere and into the corona. A prominence, like the chromosphere itself, radiates most of its light through the Hydrogen-alpha transition.

Pyrex: A trademark used for any of various types of heat-resistant and chemical-resistant glass. Pyrex glass is preferred for making telescope mirrors and lenses.

Quartz glass: Quartz cut into a shape appropriate for glass application. Often used because it does not absorb UV light.

Radio telescope: A telescope that determines the position of radio signals emanating from the celestial sphere.

Radius of curvature: The radius of curvature of a curve at a given point is the radius of the circle whose curvature would match the curve at that point. It can be found by calculating the absolute value of the reciprocal of the curvature at a point on a curve.

Ramsden eyepiece: An eyepiece consisting of two plano-convex lenses with the flat surfaces facing away from each other. Lateral color is not fully corrected because the focal plane is outside the eyepiece. This lens can be used with a reticule.

Ray tracing: A powerful method of analyzing optical systems that follows the paths of individual parallel rays to see where they end up.

Reflecting prism: A prism used to change the direction of a light ray.

Reflecting telescope: A general classification of telescopes in which the primary optical element is a mirror that reflects the incoming light, rather than a lens that refracts it. Reflecting telescopes have many advantages over refracting telescopes because the primary mirrors can be made quite large, and they require only one surface to be ground and polished. Also, they reflect all wavelengths equally and so do not suffer from chromatic aberration.

Refracting telescope: A telescope that uses a lens instead of a mirror as its primary light-gathering element.

Resistor: A device used to control current in an electric circuit by providing resistance.

Resolving power: The ability of an optical device to produce separate images of close objects.

Reticule: A grid or pattern placed in the eyepiece of an optical instrument, used to establish scale or position.

Right ascension: One of the coordinates that locates an object on the celestial sphere. It corresponds to longitude on the surface of the earth and is measured in hours, minutes, and seconds eastwards from the point where the sun's path, the ecliptic, once a year intersects the celestial equator.

Rouge: A reddish powder, mostly ferric oxide, used for polishing metals or glass.

Schmidt telescope: A wide-angle reflecting telescope used primarily for astronomical photography, in which spherical aberration and coma are reduced to a minimum by means of a spherical mirror with a corrector plate near its focus.

Scintillation: In astronomy, the rapid variation in the light of a celestial body caused by turbulence in earth's atmosphere; a twinkling.

Scintillometer: A device that measures the amount of twinkling caused by the earth's atmosphere.

Secondary mirror: A second mirror surface in a reflector or in a catadioptric telescope.

Seeing: A measure of the distortion created by the atmosphere. Although stars are point-like objects, the atmosphere blurs them into a brightness profile that is gaussian shaped. The seeing is defined to be the width of the disk that contains roughly 62 percent of a star's light.

Selenography: The study of the physical features of the moon.

Selenographer: One who studies selenography.

Silvered mirror: A glass mirror coated with a thin layer of reflective silver. Silver is almost never used today because it tarnishes quickly.

Sinusoidal: This word describes any function that has the shape of a trigonometric sine curve.

Soda lime glass: An inexpensive glass made from sand (silica), soda ash (sodium carbonate), limestone (calcium carbonate), and feldspar in quantities of approximately 56, 18, 18, 7 percent respectively.

Solar disk: The visible disk of the sun. See also **Photosphere.**

Spectrograph: A device that separates light into its component colors and records these on a photographic plate. Spectrographs reveal much about the chemistry of stars and are an indispensable tool of astronomy.

Spectrographic film: Spectrographic film is specially designed for use in spectrographs. It comes in long strips, and is relatively large grained and extremely sensitive to all colors.

Spectrohelioscope: An apparatus for viewing the sun with a monochromatic light in order to show the details of the sun's surface and surroundings as they would appear if the sun emitted only that light.

Spectrum: The band of colors (red, orange, yellow, green, blue, indigo and violet) that is produced when white light is split into its constituent wavelengths by passing it through a prism.

Speed Graphic: A specific bellows-type camera that was popular with reporters from the 1930s to the 1950s.

Spherical aberration: Variation in focal length of a lens or mirror from center to edge, due to its spherical shape.

Spherical mirror: A mirror with a spherical curvature. The spherical shape is the easiest to create in a mirror, but such mirrors do not focus parallel rays to the same point and so suffer from what's called spherical aberration.

Spotting telescope: A small telescope with a low magnification used to spot interesting features in a field for further scrutiny. (Also called a **finder** or **finderscope.**)

Sputtering: A process that uses ions of an inert gas to dislodge atoms from the surface of a crystalline material. The atoms are then electrically deposited to form an extremely thin coating on a glass, metal, plastic, or other surface.

Stellar photometry: Recording the brightness of stars.

Sunspot: A sunspot is a region in which strong magnetic fields emerge into the solar atmosphere from below the solar surface. Sunspots can be observed in both Hydrogen-alpha light and the white light continuum. Sunspots are cooler then their surroundings because the magnetic fields that form them reduce the temperatures of the plasma that they contain.

Terrestrial telescope: A telescope designed to look at objects on the earth. So as not to confuse the viewer's eye, a terrestrial telescope is erecting.

Triplet lens: A system of three lenses that is used to counteract the effects of spherical and chromatic aberration.

Ultraviolet light: The range of invisible light with wavelengths from about 4 nanometers to about 380 nanometers.

Vacuum tube: An electronic device that uses flowing electrons inside an evacuated glass tube to affect an electric signal.

Variable frequency oscillator: Any oscillator that always repeats the same pattern, but at an adjustable rate.

Variable star: A star that varies markedly in brightness over time.

Vinetting: A kind of image distortion that can occur in optical systems.

Viscosity: The degree to which a fluid resists flow under an applied force, measured by the tangential friction force per unit area divided by the velocity gradient.

Visible light: Visible light ranges from wavelengths of 400 nanometers to about 750 nanometers.

Volt: Unit of electric potential and electromotive force, equal to the difference of electric potential between two points on a conducting wire carrying a constant current of one ampere when the power dissipated between the points is one watt.

Voltage transformer: A device that uses inductance to step up or down a voltage that varies with time.

Watt: A unit of power. One watt is delivered when one joule of energy is transferred in one second.

Yagi Antenna: A radio antenna that consists of a number of parallel bars fixed perpendicularly on a long rod. This antenna is directional, but it can only pick up a narrow band of wavelengths. As a result, it is not often used these days in radio astronomy. (See **Parabolic antenna.**)

Zenith: The point on the celestial sphere that is directly above the observer.

FURTHER READING

Background Information

Astronomy Explained, North, G. Springer-Verlag, 1997 (Advanced)

Bubbles, Voids and Bumps in Time: The New Cosmology, Cornell, J. editor, Cambridge University Press, 1992 (Advanced)

Cambridge Atlas of Astronomy, Cambridge University Press, 3rd edition 1994 (Good overview of what's known)

Galactic and Extragalactic Radio Astronomy, Verschuur, G. L., and Kellermann, K. I., editors, Springer-Verlag, 1974 (Really advanced)

Encyclopedia of Astronomy and Astrophysics, Meyers, R. A., editor, Academic Press, 1989 (Good overview and great reference)

Principles of Physical Cosmology, Peebles, P. J. E., Princeton, 1993 (Only geniuses need apply)

Astronomical Calculations

Astronomical Methods and Calculations, Acker, A and Jaschek, C. John Wiley, 1986 (Beginner)—out of print

Astrophysical Formulae, Lang, K. R., Springer-Verlag, 1999 (For serious amateurs only)

Practical Astronomy with Your Calculator, Duffett-Smith, P. J. Cambridge University Press, 1989 (Beginner)

Astronomy with Your Personal Computer, Duffett-Smith, P. J. Cambridge University Press, 1990 (Beginner)

Telescope Making/Optics

Amateur Telescope Making, Volumes 1, 2 and 3, Ingalls, Albert G. Willman-Bell, 1996

Astronomical Optics, Schroeder, D. J., Academic Press, 1987 (Advanced telescope makers only, please)

How to Make a Telescope, Texereau, J. (translated by A. Strickler), Willmann-
Bell, 1984 (Good introduction)

*Star Testing Astronomical Telescopes: A Manual for Optical Evaluation and
Adjustment,* Suiter, H. R., Willmann-Bell, 1994 (Amateur accessible)

*Star Ware: The Amateur Astronomer's Ultimate Guide to Choosing, Buying,
and Using Telescopes and Accessories,* Second Edition Harrington, Philip,
John Wiley & Sons, Inc., 1998 (Amateur accessible)

Unusual Telescopes, Manly, P. Cambridge University Press, Second edition,
1995 (Amateur accessible. Delightful for any technophile with a taste
for original approaches)

Observers Guides

Advanced Amateur Astronomy, North, G., Cambridge, 1997 (First-class summary of advanced amateur projects. Highly recommended for any amateur astronomer who wants to get serious about observing)

Burnham's Celestial Handbook, Volumes 1, 2 and 3. Burnham, R., Dover
1983.

Compendium of Practical Astronomy: Volume 1—Instrumentation and
reduction techniques. Volume 2—Earth and Solar System. Volume 3—
Stars and stellar systems. Edited by Roth, G. D., Springer-Verlag, 1995
(For serious amateurs only!)

The Observer's Guide to Astronomy (2 vols.), Martinez, P. Cambridge University Press, 1994 (Amateur accessible)

*The Universe from Your Backyard: A Guide to Deep Sky Objects from
Astronomy Magazine,* Eicher, D. J. Cambridge University Press, 1988
(Wonderful!)

Visual Astronomy of the Deep Sky, Clark, R. N. Cambridge University Press,
1989 (Amateur accessible)—out of print

Astrophotography and Imaging

Astrophotography for the Amateur, Covington, M. Cambridge University
Press. Second edition, 1999 (Good reference)

CCD Astronomy: Construction and Use of an Astronomical CCD Camera,
Buil, C. Willmann-Bell, 1991 (Amateur accessible. Good reference)

Choosing and Using a CCD Camera, Berry, R., Willmann-Bell, 1992 (Amateur accessible. Solid reference)

A Manual of Advanced Celestial Photography, Wallis, B. and Provin, R.
Cambridge University Press, 1988 (Outstanding reference for serious
amateurs)—out of print

High Resolution Astrophotography (Practical Astronomy Handbook #7), Dragesco, J. Cambridge University Press, 1995 (Amateur accessible but you'll work hard reading this book)

The Sun

Eclipse! Harrington, P., John Wiley & Sons, Inc., 1998.

Guide to the Sun, Phillips, K. J. H. Cambridge University Press, 1995 (Amateur accessible)

Observe Eclipses, Reynolds M., and Sweetzin, R. Astronomical League Sales, 1995 (Amateur accessible)

Observing the Sun (Practical Astronomy Handbook #3), Taylor, P. O., Cambridge University Press, 1992 (Buy this book!)

Solar Astronomy Handbook, Beck, Hilbrecht, Reinch, Volker, Willmann-Bell, 1995 (Great reference!)

Stars, Photometry and Nebulae

Astronomical Photometry: Text and Handbook for the Advanced Amateur and Professional Astronomer, Henden, A. and Aitchuk, Willmann-Bell, 1990 (Great reference)

Getting the Measure of the Stars, Couper, W. A. and Walker E. N. Adam Hilger, 1989 (Amateur accessible)

Observing Variable Stars: A Guide for the Beginner, Levy, D. H. Cambridge University Press, 1989 (Highly recommended!)

Observing Visual Double Stars, Couteau, P. (translated by A. Batten). M.I.T. Press, 1981 (Good reference)—out of print

Photoelectric Photometry of Variable Stars. A Practical Guide for the Smaller Observatory, Hall, S. and Gennet, R. M. Willmann-Bell, 1988 (Amateur accessible if you work at it)

Planetary Nebulae—A Practical Guide and Handbook for Amateur Astronomers. Hynes, S. J., Willmann-Bell, 1991 (Good reference)

Supernova Search Charts and Handbook, Thompson, G. D. and Bryan, J. T. Cambridge University Press, 1989 (Amateur accessible, with work)

Variable Stars, Petit, M. John Wiley, 1987 (Good reference)—out of print

Advanced Techniques

Advanced Amateur Astronomy, North, G., Cambridge University Press, 1997 (I liked this book so much I decided to list it twice)

Astrophysical and Laboratory Spectroscopy, Brown R., Lang J. L., Editors, Edinburgh University Press, 1988 (Quite advanced, but it will tell you just about everything you ever wanted to know [and a few things you didn't] about spectroscopy)—out of print

Interferometry and Synthesis in Radio Astronomy, Thompson, A. R., Moran, J. M., and Swenson, G. W. John Wiley, 1986 (Really Advanced, I mean wow!)—out of print

Stars and their Spectra, Kaler, J. B., Cambridge University Press, 1989 (That's right, it's advanced)

Star Atlases and Catalogues

Uranometria 2000.0 (Vols. I & 2), Tirion, W. et al., Willmann-Bell, 1987 & 1988.

Sky Atlas 2000.0, Second Edition Tirion, W., Sinnott, R. W., Cambridge University Press, 1999 (Foundational)

Deep-Sky Companions: The Messier Objects, O'Meara, S. J. and Levy, D. H. Cambridge University Press, 1998.

The Deep-Sky Field Guide to Uranometria 2000.0., Cragin, Lucyk and Rappaport, Willmann-Bell, 1993 (Fundamental)

The Deep Space Field Plan, Vickers, J. C., Sky Publishing Corp., 1990 (Deep fielders alert! Buy this book)

NGC 2000.0. The Complete New General Catalogue (NGC) and Index Catalogue (IC) of Nebulae and Star Clusters, Edited by Sinnott, R. W., Cambridge University Press, 1989 (Foundational)—out of print

Norton's Star Atlas and Reference Handbook (Epoch 2000.0), Edited by Ridpath, I., Longman Publishing Group, 1998 (Foundational)

Sky Catalogue 2000.0, Volume 1: Stars to magnitude 8. (1982), Volume 2: Double stars, variable stars, and nonstellar objects. (1992) Both volumes edited by Hirshfield, A. and Sinnott, R. W., Cambridge University Press (Foundational)

The AAVSO Variable Star Atlas, Scovil, C. E. Sky Publishing Corporation, 1980 (If you want to observe variable stars, this is an indispensable reference)

Magazines

Sky & Telescope
> The best all around magazine for amateur astronomers who want to learn the how-to secrets of observational astronomy. If you have even a passing interest in hands-on work, you should subscribe to this publication. Check your newsstand or write to

Sky Publishing Corporation; 49 Baystate Road, Cambridge, MA 02138

Astronomy Magazine
> Visually stunning. This publication is great for the dabbler who enjoys learning and looking at fantastic images of exotic objects. Lots of great

information and stunning stellar mug shots. But the hands-on doer will prefer *Sky and Telescope.* Check your local magazine racks to see if you agree.

Scientific American

Need we say more. For subscriptions, write or call Scientific American, 415 Madison Ave., New York, NY 10017; 1-800-333-1199. You may also want to check out their electronic archive at *www.sciamarchive.com.*

Editor's Note: Many of the wonderful books listed are no longer in print. But with the power of the internet, it is now easier than ever to connect with an owner who may be willing to part with his or her copy. Or you may want to try one of the online bookstores that deal in out-of-print books, such as *www.alibris.com.*

CONTACT LIST

Supplier List

Celestron. 2835 Columbia St., Torrance, CA 90503. 310-328-9560 *www.celestron.com* Telescopes and accessories.

Edmund Scientific. 101 E. Gloucester Pike, Barrington, NJ 08007-1380. 800-728-6999 *www.scientificsonline.com*

Hamamatsu. 360 Foothill Rd., Bridgewater, NJ 08807. 732-356-1203 *www.hamamatsu.com* Photomultiplier tubes, power supplies, counters and electronics

Lumicon Corporation. 2111 Research Dr. No. 5, Livermore, CA 94550. 925-447-9570 *www.lumicon.com* Telescopes, filters, and accessories

Marconi Applied Technologies. 4 Westchester Plaza, PO Box 1482, Elmsford, NY 10523. 914-592-6050 *www.marconi-technologies.co.uk* CCDs and CCD cameras

Meade Instruments. 6001 Oak Canyon, Irvine, CA 92618. 949-451-1450 *www.meade.com* Telescopes, components and accessories.

Optec, Inc. 199 Smith Street, Lowell, MI 49331. 616-897-9351 *www.optecinc.com* Photometric equipment, including the SSP series photometers

Orion Telescope Center, Box 1815-S Santa Cruz, CA 95061-1815. 800-676-1343 *www.telescope.com* Telescopes, eyepieces and other and accessories, as well as binoculars.

Photometrics Ltd. 3440 East Britannia Drive, Tucson, AZ 85706. 520-889-9933 *www.photomet.com* CCD cameras and equipment.

Santa Barbara Instrument Group. P.O. Box 50437, 1482 East Valley Road No. 33, Santa Barbara, CA 93150. 805-969-1851 *www.sbig.com* Makers of the SBig CCDs.

Spectra Astro Systems. 20620 Lassen St., Chatsworth, CA 91311. 800-735-1352 *www.spectraastro.com*

Tele Vue Optics. 100 Route 59, Suffern, NY 10901. 914-357-9522
www.televue.com Refracting telescopes, eyepieces and accessories.

Texas Instruments. P.O. Box 655303 MS 8206 Dallas, TX 75265.
972-995-2011 *www.ti.com* CCDs and CCD cameras

Tinkers Guild. 405 El Camino Real PMB 326, Menlo Park, CA 94025.
650-853-1001 *www.tinkersguild.com* Books, telescopes and accessories

Thousand Oaks Optical. Box 4813, Thousand Oaks, CA 91359
805-491-3642 *www.thousandoaksoptical.com* Full-aperture solar
filters and eclipse viewers

University Optics, Ann Arbor, MI

Organizations

American Astronomical Society. 2000 Florida Ave.—Suite 400, NW,
Washington DC 20009-1231. 202-328-2010 *www.aas.org*

Astronomical Society of the Pacific. 390 Ashton Ave., San Francisco, CA
94112. 415-337-1100 *www.asp-sky.org*

American Association of Variable Star Observers. 25 Birch St., Cambridge, MA 02138. 617-354-0484 *www.aavso.org*

Association of Lunar and Planetary Observers. Harry Jamieson, ALPO
Membership Secretary, PO Box 171302, Memphis, TN 38187-1302
www.lpl.arizona.edu/alpo/

Society for Amateur Scientists, 4735 Clairemont Square PMB 179, San
Diego, CA 92117 *www.sas.org*

Southern California Observatory for Public Education (SCOPE), 13201
Boy Scout Camp Road, Lockwood Valley, CA 93225 661-245-3867
www.socal-observatory.org/

INDEX